Also of interest

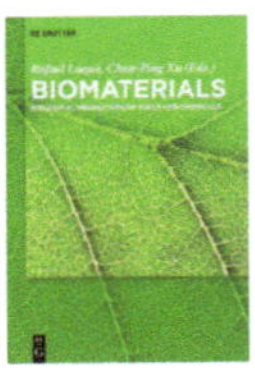

Biomaterials.
Biological Production of Fuels and Chemicals
Editors: Rafael Luque and Chun-Ping Xu, 2016
ISBN 9783110342307, e-ISBN (PDF) 9783110342420

Photocatalysis
Editors: Suresh C. Pillai and Vignesh Kumaravel, 2021
ISBN 9783110668452, e-ISBN (PDF) 99783110668483

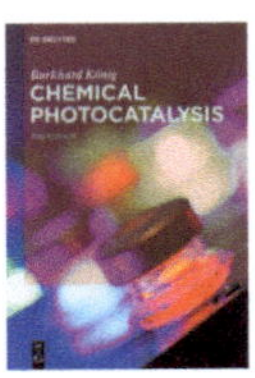

Chemical Photocatalysis
Burkhard König, 2020
ISBN 9783110576542, e-ISBN (PDF) 9783110576764

Nanomaterials.
Volume 1: Electronic Properties
Engg Kamakhya Prasad Ghatakaund Madhuchhanda Mitra, 2018
ISBN 9783110609226, e-ISBN (PDF) 9783110610819

Nanomaterials.
Volume 2: Quantization and Entropy
Engg Kamakhya Prasad Ghatak and Madhuchhanda Mitra, 2020
ISBN 9783110659726, e-ISBN (PDF) 9783110661194

Biomaterials for Photocatalysis

Promising New Materials

Edited by
Awais Ahmad, Hina Sharif, Rafael Luque,
Mabkhoot Alsaiari and Farid A. Harraz

DE GRUYTER

Editors
Awais Ahmed
Departamento de Química Orgánica
Universidad de Córdoba
Campus de Rabanales
Carretera Nacional IV-A, km 396
14014 Córdoba
Spain
awaisahmed@gcuf.edu.pk

Dr. Hina Sharif
Department of Pharmaceutical Chemistry
Government College University Faisalabad
Punjab 38000
Pakistan
hina.sharif17@gmail.com

Prof. Rafael Luque
Departamento de Química Orgánica
Universidad de Córdoba
Campus de Rabanales
Carretera Nacional IV-A, km 396
14014 Córdoba
Spain
q62alsor@uco.es

Prof. Mabkhoot Alsaiari
Advanced Materials and Nano-Research Centre
Najran University (PCSED)
11001 Najran
Saudi Arabia
Mabkhoot.Alsaiari@gmail.com

Prof. Farid A. Harraz
Advanced Materials and Nano-Research Centre
Najran University (PCSED)
11001 Najran
Saudi Arabia
faharraz@nu.edu.sa

ISBN 978-3-11-076871-8
e-ISBN (PDF) 978-3-11-076874-9
e-ISBN (EPUB) 978-3-11-076883-1

Library of Congress Control Number: 2022950031

Bibliographic information published by the Deutsche Nationalbibliothek
The Deutsche Nationalbibliothek lists this publication in the Deutsche Nationalbibliografie; detailed bibliographic data are available on the internet at http://dnb.dnb.de.

Cover image: greenleaf123/iStock/Getty Images
Typesetting: Integra Software Services Pvt. Ltd.
Printing and binding: CPI books GmbH, Leck

www.degruyter.com

Preface

Though synthetic polymers are commonly used photocatalysts, researchers were urged to develop sustainable and eco-friendly biomaterials using different biopolymers such as agar, alginate, chitin, cellulose, dextran and derivatives, gelatin, gums and collagen due to their multiple applications and abundance in nature. Their natural properties such as large surface area, nontoxic nature, antibacterial activity, biocompatibility and ease of handling have made them important. Different approaches such as sol–gel method, ultrasonication, green synthesis, microwave-assisted synthesis, microbial synthesis and electrospinning techniques have been fabricated to make biopolymers for photocatalysis applications such as biohybrids, composite films, photocatalytic aerogels, nanocomposites and heterojunctions. Surface functionalization and deacetylation are used to obtain derivatives of biopolymers such as deacetylation of chitin biopolymers to form chitosan. To make hydroxyapatite, fish bone is pretreated by washing and boiling with acetone, and then subjected to hydrolysis and calcination treatment. These biomaterials are really promising materials which scale up their production by fascinating results in different catalytic applications.

Awais Ahmad,
Hina Sharif,
Rafael Luque,
Mabkhoot Alsaiari and
Farid A. Harraz

https://doi.org/10.1515/9783110768749-202

Contents

List of authors

Awais Ahmad
Departamento de Quimica Organica
Universidad de Cordoba
Edificio Marie Curie (C-3)
Ctra Nnal IV-A, km 396
E14014 Cordoba
Spain
awaisahmed@gcuf.edu.pk
Chapters 1, 2, 3, 4, 5, 6, 7, 8, 9, 10

Prof. Mabkhoot Alsaiari
Promising Centre for Sensors and Electronic Devices (PCSED)
Advanced Materials and Nano-Research Centre
Najran University
Najran 11001
Saudi Arabia
and
Empty Quarter Research Unit
Department of Chemistry
College of Science and Art in Sharurah
Najran University
Sharurah
Saudi Arabia
Mabkhoot.Alsaiari@gmail.com
Chapters 1, 2, 3, 4, 5, 6, 7, 9, 10

Prof. Farid A. Harraz
Promising Centre for Sensors and Electronic Devices (PCSED)
Advanced Materials and Nano-Research Centre
Najran University
Najran 11001
Saudi Arabia
and
Nanomaterials and Nanotechnology Department
Central Metallurgical Research and Development Institute (CMRDI)
P.O. 87 Helwan
Cairo 11421
Egypt
faharraz@nu.edu.sa
Chapters 1, 2, 3, 4, 5, 6, 7, 9, 10

Prof. Rafael Luque
Departamento de Quimica Organica
Universidad de Cordoba
Edificio Marie Curie (C-3)
Ctra Nnal IV-A, km 396
E14014 Cordoba
Spain
q62alsor@uco.es
Chapters 1, 2, 3, 4, 5, 6, 7, 8, 9, 10

Arsh-e-Noor
Department of Environmental Science and Engineering
Government College University Faisalabad
Punjab 38000
Pakistan
arshenoor164@gmail.com
Chapter 8

Dr. Hina Sharif
Department of Pharmacy
The University of Faisalabad
Punjab 38000
Pakistan
hina.sharif17@gmail.com
Chapters 1, 6, 7, 8, 9, 10

Sadaf Tariq
Department of Biochemistry,
Government College University Faisalabad
Punjab 38000
Pakistan
sadaftariq787@gmail.com
Chapters 2, 3, 4, 5

https://doi.org/10.1515/9783110768749-204

Awais Ahmad, Hina Sharif, Rafael Luque, Mabkhoot Alsaiari,
Farid A. Harraz

1 Biomaterials for photocatalysis: an overview

Abstract: Light-responsive bioinspired materials are emerging materials for precise and controllable medical and biological applications. These light-responsive bioinspired materials have been studied to design light- induced modulators, nanovehicles and biopolymers to control cell behavior, as well as regulate environments. Photocatalysis uses sustainable, eco-friendly, natural light and power for effective chemical conversion. Biomaterial-assisted photocatalysis has become a hot spot for research due to its applications: degradation of pollutants in water and air, artificial photosynthesis, photochromism, cancer therapy, drug delivery, dissipation of heat and energy, sterilization and production of affordable energy sources. Photo-generated electrons, bandgap energy and generation of electron–hole pair are applied to degrade or create different compounds by absorbing or releasing energy or heat. This chapter will focus on the principle of photocatalysis along with biomaterial-assisted photocatalysts.

Keywords: Photocatalysts, biomaterials, nanochemistry, enzyme, biotransformation, reduction, oxidation

1.1 Introduction

A combination of light-responsive surfaces and nanoparticles has emerged as an innovation that provides positive sustainability to society. Biomaterials designed by mimicking natural materials have become a current topic for research. The idea to form biomaterial-assisted photocatalysts originated from photosynthesis and became an effective option to produce renewable-energy [1]. Such photocatalysts have gained much importance due to their applications in removal of pollutants from water and environment. Photocatalysts can be activated through both UV and visible light. When photocatalysts are biomaterial-based, they provide greater surface area, better functionality and porosity for not only better dispersion but also increased interaction with contaminants. Biomaterial-assisted photocatalysis uses natural materials to increase the compatibility of this approach with living systems [2].

For ideal photocatalysis, there should be strong interaction, larger surface area, greater adsorption, easy separation and higher resistance to denaturation of catalyst. The biomaterials can be biopolymers, biochars, biomolecules, immobilized

https://doi.org/10.1515/9783110768749-001

enzymes, proteins or peptides for excellent photocatalytic activity [3]. Usually, inorganic photocatalysts have better activity than organic, due to optimum bandgaps and greater electron flow. Moreover, biomaterial-based photocatalysts are preferred over inorganic photocatalysts due to their environment-friendly nature and lowest particle size [4, 5].

Various factors such as improved light absorption and greater photon harvesting capability that can be related to size and structure of immobilized biomaterials and crystallinity [6, 7] influence the efficiency of photocatalytic activity. As regards physiochemical properties, the interactions and biomaterials are very important. During typical photocatalysis, photons are absorbed at the interface, and thereby, an electron–hole pair is generated to form hydroxide and superoxide anion radicals [8]. Electrons generated in light-responsive reactions undergo reduction reactions, and the charge carriers are involved in photocatalysis [9, 10]. Electrons play their role in reduction reactions, whereas in reduction of pollutants, holes are involved in oxidation reactions; they oxidize pollutants [11, 12].

The bandgap energy plays an important role in formation of free radicals to start redox reactions. Various bottlenecks can affect the efficiency, such as spectral mismatch, recombination of photogenerated electrons and low surface area [13, 14]. Therefore, different strategies have been developed to enhance the biomaterial-assisted photocatalysis under UV or visible light, so far [15, 16], by tuning the bandgap according to natural spectra to avoid spectral mismatch either by doping, modifying bandgaps or sensitization [17, 18].

Hence, biomaterials with engineered nanoparticle-based photocatalysts are used in various applications including detoxification, packaging and to combat environmental pollution [19, 20]. In subsequent sections of this chapter, the fundamental principle of photocatalysis will be discussed, followed by an overview of biomaterial-based photocatalysts as a presentation of recent discovery in the field of biomaterials. Different modes of synthesis and techniques to characterize biomaterials for photocatalysis will also be discussed. Challenges in modulation along with strategies to develop an efficient photocatalyst with future directions will be provided as a critical area for biomaterials.

1.2 Basic principle of photocatalysis

Photocatalysis involves light and a catalyst to proceed with a chemical reaction that involves valence band (filled) and conduction band (empty) with energy gaps (bandgap measured as electron volt, eV) [21]. Any photon having bandgap energy can activate an electron from ground state (valence band) to jump to excited state (conduction band) – an essential step during photocatalysis because electron of conduction band and the hole in the valence band tends to move in lattice [22].

These free electrons and holes are involved in ongoing and repeated redox reactions during photocatalysis. The redox reactions are useful either to remove toxic ions by dissolving them or mineralize dissolved toxic ions. Photocatalysis can be done through different mechanisms, based on the incident wavelength of light on the surface [23].

1.2.1 When the wavelength of light is >400 nm

Mostly, when dyes absorb visible light, photocatalytic degradation occurs through direct mechanism. In direct mechanism, a photon excites the electron from its ground state to excited state, thereby converting it to a semi-oxidized racial cation through electron transfer in conduction band. The trapped electrons tend to react with oxygen and form superoxide radical anion, converting it to a hydroxyl radical; this radical further undergoes oxidation reaction [24].

1.2.2 When the wavelength of light is <400 nm

1.2.2.1 Generation of electron–hole pair

When the wavelength of light is <400 nm, the electrons does not move into excited state; in that case, the photocatalyst helps the reaction to proceed. The bandgap is typically higher, and the available energy is greater than the bandgap. Hence, the photons from UV light tend to promote the movement of a photoelectron from valence band of a photocatalyst to conduction band, leaving a hole behind in valence band to generate an electron–hole pair [6].

Ionization of water

The holes generated at the valence band, in response to UV light, react with water to form hydroxide radical on the surface of a photocatalyst. It causes biomineralization of organic molecules by reacting at the surface of a photocatalyst. The extent of biomineralization depends upon the structure and stability of the respective organic molecule [20].

Sorption of oxygen ions

The electrons of the conduction band react with oxygen ions and generate anionic superoxide radicals, leading to an oxidation reaction, and prevent the generation of electron–hole pair, thus retaining the neutral nature within the molecule.

1.2.2.2 Protonation of superoxide

Apart from generation of an electron–hole pair, protonation of superoxides by forming hydroperoxyl radicals is another approach to photocatalysis. The hydroperoxyl radical produces hydrogen peroxide, further dissociates, and forms hydroxyl radicals, reacting with pollutants to form carbon dioxide and water [17].

1.3 Overview of biomaterials for photocatalysis

1.3.1 Biopolymers-based photocatalysts

Though synthetic polymers are commonly used photocatalysts, researchers were urged to develop sustainable and eco-friendly biomaterials using different biopolymers such as agar, alginate, chitin, cellulose, dextran and derivatives, gelatin, gums and collagen due to their multiple applications and abundance in nature. Their natural properties such as large surface area, nontoxic nature, antibacterial activity, biocompatibility and ease of handling have made them important [23]. Different approaches such as sol–gel method, ultrasonication, green synthesis, microwave-assisted synthesis, microbial synthesis and electrospinning techniques have been fabricated to make biopolymers for photocatalysis applications such as biohybrids, composite films, photocatalytic aerogels, nanocomposites and heterojunctions. Surface functionalization and deacetylation are used to obtain derivatives of biopolymers such as deacetylation of chitin biopolymers to form chitosan. To make hydroxyapatite, fish bone is pretreated by washing and boiling with acetone and then subjected to hydrolysis and calcination treatment [21].

Biopolymers are used as photocatalytic materials for various applications such as decomposition of pollutants in contaminated water due to (i) abundance: biopolymers are bioinspired materials and can be extracted either from natural sources as chitin can be extracted from crabs, shell of shrimps and lobsters or from chemical modification of derived biopolymers as chitosan from deacetylation of chitin biopolymer; (ii) high functionality: biopolymers can perform a variety of functions such as binding pollutants through their chemical bonds to remove pollution; and (iii) nontoxicity: natural and derived biopolymers are nontoxic to living systems [24].

Cai [24] synthesized TiO_2/chitin nanohybrid for photocatalytic activity, by using methylene blue. Zhang et al. synthesized chitin whisker Ag_3PO_4 biocomposites for photocatalytic degradation of rhodamine B dye, by chemical deposition in 15 min. Chitin was found to act as an electron transfer channel to favor photogenerated electrons transfer at the surface by dispersing electrons and making them stable photocatalysts. Another photocatalyst, Cu_2O-chitin ternary biocomposite has been reported in decomposition of methyl orange dye by scattering electrons, by improving photodegradation and decreasing rate of formation of electron–hole pairs [21]. Both metal-based photocatalysts and metal- free photocatalysts can be composited on polymer surfaces for photocatalytic degradation of pollutants as chitin-derived carbon/g-C_3N_4 heterojunctions [14].

1.3.2 Biochar-based photocatalysts

Carbon-rich porous solid biochars produced by thermal decomposition of biomass are commonly used in synthesis of photocatalysts for their low cost, larger surface area and strong conductivity. Recently, biochars have been used as a support for TiO_2 to promote its photodegradation and catalytic efficiency by using synergistic effect, which enabled an immediate reaction between pollutants and the TiO_2 surface. Biochar-based photocatalysts can be formed through incorporation of photocatalytic nanoparticles to biochar templates for low cost, better surface properties, improved photocatalytic competency, better crystallinity and higher chemical stability than simple photocatalysts. Likewise, the biochar-based photocatalysts have improved charge carrier density, bandgap and half-life of photo-charge separation by the sol–gel method and hydrolysis method [23]. Biochars have been found to act as electron-acceptors by decreasing the rate of electron–hole pair formations, promoting its e^-–h^+ separation, increasing synthesis of OH•, increasing spectral absorbance and improving photocatalytic activity. A similar approach was followed by [22] using ZnO as photocatalyst. Notably, the activity of the photocatalyst can be improved by addition of nitrogen-containing functional groups in nitrogen-doped biochars for better electron transport efficiency and improved photochemical properties. TiO_2 biochars are used as highly efficient photocatalysts for photocatalytic degradation [25].

For improving the photocatalytic performance, a biochar can be used as it has low cost and is an easily available carbon material for improving the visible light absorption, making available photogenerated carriers and separation of electron–hole pairs in response to light, such as biochar-BiOX composite biomaterials. Also, to reuse any catalyst is also very vital for photocatalytical application of biochar. Fe_3O_4 can be used to recycle photocatalyst; for instance, Fe_3O_4/BiOX/biochar was successfully recycled [22], with up to 90% recycle capacity.

Biochars can be used to support nanoparticles by tuning the functionality and pollutant removal competency via adsorption or photocatalysis through shuttle electrons, by improving electrical conductivity, increasing exposed surface area, improving charge separation, acting as electron reservoir and reducing bandgap energy [23, 26].

1.3.3 Biomolecule-based photocatalysts

Many biomaterials can be used to assist the synthesis of photocatalysts due to their special biocompatible and biochemical nature, making them attractive for use in artificial methods for future applications. The synthesis of biomaterial-assisted photocatalysts can be done in two ways: (i) biometrics with different size ranges like leaves, viruses or wings; and (ii) secreted, extracted or inspired biomolecules from plants and microorganisms, such as silk, chitin, chitosan, biosilica, DNA and proteins. Biomolecules assist the synthesis by fabrication of plasmonic photocatalysts inspired by metal nanoparticles as key components. Generally, the synthesis is done through in situ method. Liu et al. [27] synthesized gold and silver nanoparticles-assisted photocatalysts by exploiting the microfluids in Magnolia leaves through conversion of *p*-nitrophenol into *p*-aminophenol. The solution-based method uses toxic compounds as precursors; these toxic compounds are adsorbed and reduced into metals by biomass and the living system [10].

Though synthesis of nanoparticles happens in harsh conditions, biomolecule-assisted synthesis takes place at ambient environmental conditions, without affecting environment. Liu et al. [27] have reported synthesis of plasmonic silver photocatalyst under mild conditions through lysogeny broth. $AgNO_3$ and NaCl underwent a precipitation reaction to form silver chloride nanospheres. Silver nanoparticles were synthesized by irradiation on silver chloride surface, which generated electrons. The time of irradiation and the concentration used during the reaction can fine-tune the bandgap [28]. These were found to show excellent oxidizing properties against pollutants and microorganisms. These advanced photocatalysts are now being used to address different environmental demands like energy conversion and green chemistry [29].

Mussels are unrestrained fouling organisms, having classic adherent properties to attach with all types of surfaces, either organic or inorganic, and even with adhesion-resistant surfaces such as poly(tetrafluoroethylene). This versatile nature of mussels is due to the presence of specific proteins near its interface; 3,4-dihydroxy-L-phenylalanine amino acid is mostly common in these proteins. Polydopamine (PDA) coating is another adhesive protein in mussels. This approach was adopted in the synthesis of biomaterials; the coatings on the surface were fabricated on different supports such as silica beads and PET films. This strategy enhanced the conversion of L-glutamate, thereby coupling the enzyme with photochemical regeneration of NADH [1, 23].

1.3.4 Immobilized enzymes and peptide-assisted photocatalysts

For different biocatalytic applications, immobilized enzymes and peptides can be used, as in environmental detoxification. However, cost-effectiveness, complexity and suitability have been real concerns. Different immobilizations can be done by crosslinking, adsorption and entrapment. Optimal techniques are also important to preserve activity and stability of immobilized enzyme (Table 1.1).

For example, peroxidase is known for its biocatalytic activity, high stability and broad substrate specificity. Di Paola [30] prepared a photocatalyst using Ti and peroxidase (TiO_2-peroxidase) by integrating into matrix. Soybean peroxidase was used for its broad range of substrate specificity. Additionally, peptides and proteins are also used in photocatalytic systems. Hemin and iron-porphyrin that are comprised of π-conjugated system show biomimetic catalytic activity.

Chen et al. made imidazole-coordinated hemin for photocatalysis of phenols and other antibiotics. This hemin catalytic coordination showed an excellent photocatalysis of 4-chlorophenol in comparison to intact hemin, as imidazole ligands helped the peroxide cleavage by donating electrons leads to form Iron oxide. Under visible light, reactive oxygen species were formed by generation of electron–hole pair.

Biomimetic photocatalysts, by using hemin-coordinated bismuth tungstate, were formed by immobilization of hemin and tungstate layer for antibiotic (tetracycline) degradation. During photocatalysis, hemin acted as electron shuttle to transfer oxygen [21].

Hemin can also be used to make TiO_2/GO/hemin catalyst through sol–gel method to degrade rhodamine B under UV–visible light. Its high efficiency over a wide pH range makes it a heterogeneous photocatalyst to generate free hydroxyl radicals.

Peptides are used as templates to immobilize photocatalysts. Platinum nanoparticles can be immobilized on graphene oxide by using oligopeptide β-sheets as a cross-linker for metal-binding thiol and amino group, to form a bridge between Pt nanoparticles and graphene oxide nanosheets. It shows excellent photocatalysis in dye degradation through electron transfer from GO to Pt, thereby transferring the hole from oxidized GO.

Peptides can also be used for fabrication. Li et al. fabricated TiO_2 to make biotemplates by self-assembling peptides to control catalytic activity of the prepared composites. The peptide-TiO_2 showed photocatalysis by using three peptides namely; HG12, HG6 and CN225. Effective photocatalysis was shown by Cu_2O, which photodegraded maximum methyl orange under visible light [31].

Table 1.1: An overview of biomolecule-assisted photocatalysts, sources, derived catalyst and activity.

Source	Derived catalyst	Activity	Reference
Biochar	Fe_3O_4/BiOBr/biochar	Significant photocatalysis of carbamazepine under UV light, permits entry of reactants in spheres by improving the electron transfer	[28]
Biochar	TiO_2/biochar	Photodecolorization and biomineralization of methyl orange dye	[24]
Biopolymer	Cellulose/titania-based nanocomposite films	High photobleaching of methyl orange	[32]
Biopolymer	Bi_2WO_6-TiO_2/starch (BT/starch)	Photodegradation of ethylene	[33]
Biochar	Fe-Mn/binary oxides-biochar (photo-Fenton system)	Photodegradation of naphthalene	[34]
Enzyme	Lactase-immobilized bacterial cellulose/TiO_2 functionalized composite membranes	Photocatalysis of reactive red X-3B dye	[35]
Protein	CG-400 resin-supported FePz (hmdtn)4	Photodegradation of Rhb	[36]
Peptide	Cuprous oxide nanostructures tuned by histidine-containing peptides	Photodegradation of dyes	[37]
Biomaterial	GC/GO/TiO_2IOSL/GOD	Photoreduction and photooxidation of hydrogen peroxide	[38]
Biotemplate	Molecular imprinted polymer (MIP) coated P25 TiO_2 photocatalyst (DNP–P25)	Photocatalysis of pentachlorophenol	[39]
Biomolecule	GOx/TiO_2/PU(glucose oxidase immobilization)	Photodegradation of AO7	[40]
Biochar	TiO_2 supported on SM biochar	Larger surface area to catalyze the degradation of orange dye	[41]
Biopolymer	CdS/chitosan composite by biomineralization process	Photodegradation of chromium red dye	[42]

1.4 Techniques to characterize biomaterials for photocatalysts

Different methods have been used to characterize physical, chemical and optical properties of biomaterials for photocatalysts through X-ray photoelectron spectroscopy (XPS), Fourier transform infrared (FT-IR), Raman spectroscopy, scanning electron microscopy (SEM) and nuclear magnetic resonance (NMR) by evaluation of its morphology, surface area, stability, chemical composition, functional groups and other chemical structures (Table 1.2).

1.4.1 Biological analysis

SEM and transmission electron microscopy (TEM) are exclusively used to analyze the morphology, surface area, crystallinity and structures related to the biological and microscopic topology of biomaterials by imaging. Xiao et al. [43] studied the structures of biochar-assisted photocatalysts and nanocrystals by SEM imaging to isolate and quantify different elements in active biochars. Othman et al. [44] studied the structure of transparent aggregated nanosheets through TEM images. Zhao et al. [31] used SEM to evaluate the biological properties of active elements in rapeseed biochars and revealed that O content decreased, and C content increased, as the temperature was raised. Other properties such as surface area, porosity and adsorption can also be evaluated by studying the adsorption-desorption isotherm. The porosity properties vary extremely with change in temperature. For instance, the biochars that have pores like a honeycomb can be broken at pyrolysis temperature 700 °C and maintained at 200 °C. Commonly, biological characterization has facilitated the study of adsorption of different pollutants to abate environmental pollution by biomaterials [45, 46].

1.4.2 Chemical analysis

X-ray diffraction (XRD), XPS, FT-IR and Raman spectroscopy are commonly used to study structures, chemical composition and conformation of biomaterials. The crystal structure of an intact and photocatalytically degraded biomaterial were studied and compared. [g-MoS2/PGBC] through XRD spectrum. Abdel-Wahab et al. [46] used XRD spectrum to investigate macroalgae and TiO_2 biocomposites. XRD spectrum was used to detect the activated carbon structure and other linked structures. Moreover, the peaks in the spectrum showed the active chemical properties of TiO_2 by indicating its good intercalation at layers. FT-IR spectrum can be used to differentiate not only the active functional groups but also the degree of carbonization of different bioinspired materials for photocatalysis. For example, the higher wave numbers at 710 cm^{-1} and stretching vibration at 875–1,035 cm^{-1}, found to be associated with increase of TiO_2, indicated the properties of T–OC– bond and charge transfer. FT-IR can be used to investigate molecular components along with degree of carbonization. XPS is used to analyze the chemical composition and surface chemistry of biomaterials. Moreover, the bond energies and elemental analysis can be done by studying the surface activity of functional groups [46].

Raman spectroscopy is used to characterize nanostructures by demonstrating the crystallinity, intensity ratio, graphitization and electronic conduction of bioinspired materials for photocatalysis. NMR is an additional spectroscopic technique used for in-depth investigations of the chemical structures. The functional groups, reactivities and stability of bioinspired materials for photocatalysis can be studied at different temperatures and can be compared by NMR spectroscopy. It provides a detailed data on not only the functional groups and hydrocarbons but also on other than C-containing functional groups, making it more suitable for analyzing bioinspired materials. Henceforth, most researchers would prefer the use of NMR and FT-IR to study chemical structure and other functional groups [47, 48].

1.4.3 Optical properties

Diffuse reflectance spectroscopy (DRS), by using UV–visible rays, is frequently used to investigate the optical properties, absorbance, energy band properties, bandgap values and photocatalysis of biopolymers, biomolecules, biochar and other bioinspired materials. Carbon content in the biomaterials expands the absorbance range

Table 1.2: Overview of characterization of biomaterials for photocatalysts.

Characterization of biomaterials for photocatalysis	**Biological properties**	Morphology (SEM, TEM, AFM)
		Surface area (SEM-EDS)
		Porosity, crystallinity and graphitization (adsorption–desorption isotherms)
	Chemical properties	Crystals (XRD, SEM, UV–visible spectroscopy)
		Electronic surface chemistry (XPS, NMR)
		Functional groups (Raman spectroscopy, FT-IR)
		Unknown and known functional groups and their structures (NMR)
	Optical properties	Energy gaps (DRS)
		Thermal activity (TGA)

and shows a redshift toward longer wavelengths. Nevertheless, it quantifies the carbon content by showing relative absorbance waves in visible region and UV region. The energy gap can be studied using the Kubelka–Munk function used to display its photocatalytic activity. Other analysis such as thermogravimetric analysis (TGA) can be used to evaluate the mass loss and relative stability. Moctezuma et al. used TGA method for analysis of the thermal stability of sulfonated biomaterial.

1.5 Mode of synthesis

Different methods such as in situ method, ultrasound-assisted synthesis, calcination, hydrothermal approach and sol–gel method have been reported for synthesis of bioinspired materials for photocatalysis.

1.5.1 In situ method

Biomaterials-assisted photocatalysts, metal oxides and biochar can be prepared by following the in situ method. Ariza et al. proposed the method to prepare TiO_2 suspensions in bacterial matrix. The bacterial cells were suspended in acidic solution, and Tio_2 was added drop by drop, followed by constant stirring. It was filtered and dried in a vacuum evaporator to obtain biomaterial- assisted photocatalysis. Ex situ method can be also followed by comparison of different BCPs. The photocatalytic hydrogels were prepared both in situ and ex situ. The bacterial hydrogels were mixed with P25, sol–gel was obtained, fabricated, mixed with deionized water, stirred, washed to remove excessive solvent and preserved. Du et al. prepared BCPs by in situ precipitation method and demonstrated its effectiveness and high performance compared to other techniques.

1.5.2 Ultrasonication

Ultrasonication or green technology has been potentially used to fabricate photocatalysts at normal conditions and yields highly pure and uniform particle coatings, instantly. Different biomaterials can be synthesized using green technology assisted by ultrasound. TiO_2-BCPs and its derivatives were prepared through this technology. Biochar can be synthesized by pyrolysis of biomass, fabricated and mixed with metal oxides, followed by sonication. The concentrated form can be obtained by evaporating the solvent in a vacuum evaporator, followed by calcination to obtain BCPs [45].

1.5.3 Sol–gel method

The sol–gel method has been used to prepare highly pure and uniform particles. Usually, the sol is synthesized by uniform mixing, followed by ageing to form gel. The shape of the nanoparticle cannot be controlled in this approach. Zhang et al. fabricated BC-TiO_2 by using coconut shell as biomass through sol–gel. The sol–gel method takes place in the following sequence: (i) ethanol is mixed with butyl titanate to make solution 1; (ii) ethanol is mixed with water, acetic acid and polyethylene glycol to make solution 2; and (iii) solution 2 is added dropwise into solution 1 and kept for 24 h to get aged. The prepared sol–gel is aged and calcinated to obtain the final product.

1.5.4 Calcination

Typically, any natural source is used to prepare biocomposite of photocatalysis activity, crushed into powder form, added into tetrabutyl titanate and potassium hydroxide, stirred well, centrifuged, washed and dried to obtain precursor. The precursor is further calcinated under inert gas at high temperature. Wang et al. fabricated bamboo to make BCP composites by calcination. The crystallinity and photocatalytic activity were enhanced and improved under both UV and visible region. Double calcination is also followed by [49] to improve the photocatalysis and remarkable stability of support.

1.5.5 Hydrothermal approach

A crystallization process involving low temperature and high pressure, hydrothermal approach, has gained success for better purity, low agglomeration, low cost, higher dispersion and other outstanding advantages. Different types of bioinspired photocatalysts have been prepared using the hydrothermal approach. Li et al. reported the synthesis of biocomposite through hydrothermal method. Waste biomass was used to prepare ZnO biocomposite through physical activation under low temperature. The results were found to show excellent photocatalytic activity, reduced bandgap and recyclability, both under UV and visible radiations.

In another study, Primo et al. synthesized TiO_2 biocomposite for photocatalysis under both/either UV and/or visible light, through multistep hydrothermal reactions. The activation of biomass was done by using appropriate agents. The activated biomass was suspended in ethanol at 27 °C, $Ti(OBu)_4$ was diluted, and added dropwise into a suspension of biomass. Afterwards, it was homogenized well; the Ti precursor was hydrolyzed by dropwise addition of deionized water. It was stirred, heated at low temperature, washed and dried overnight through the hydrothermal route. The obtained TiO_2/PCs was found to show better photocatalytic activity at 10 nm particle size, under sunlight.

1.6 Effect of various factors on photocatalysis

Various bioinspired materials for photocatalysis have been designed to obtain required properties such as stability, adsorption, and activity under UV and sunlight. The bioinspired materials can increase the photocatalysis and yield improved dispersion and high exposure to light, along with reactants aimed for various applications, notably, degradation and detoxification of pollutants of any sort [17, 47]. Different factors affect the photodegradation efficiency and rate of reaction:

(i) Nature of photocatalyst: Different photocatalysts have different photocatalytic efficiency based on their particle size, morphology, nature, surface area, purity and lattice structure. These parameters control the behavior of a catalyst, rate of adsorption, generation of electron–hole pair, and half-life of an excited electron. For instance, Titanium dioxide (TiO_2) has three different natures: brookite, anatase and rutile. Anatase is the most effective of all forms for the photocatalysis, due to stable peroxide groups at the surface. The mixed forms of anatase and rutile have been found to perform better activity than these forms used alone. Moreover, higher the exposed surface area of the photocatalyst, the higher will be the available active sites to support the reaction. It can be related to its photodegradation activity. More active sites tend to adsorb more molecules for catalysis and therefore, it increases its photodegradation activity [27].

(ii) Effect of pH: The photocatalytic activity can be regulated by controlling the pH of system, as it influences the reaction rate, directly. The adsorption of molecules on the surface of any catalyst is a key step to start the reaction; the rate of adsorption depends on the pH. Hence, pH controls the photocatalytic degradation. At low pH, holes act as oxidation species, while at higher pH, hydroxyl radicals tend to act as oxidation species. Different photocatalysts show different activities at different pH levels. Their performance can be improved by changing the pH. Some catalysts show enhanced activity at low pH, due to increased electrostatic attraction; increased adsorption leads to better catalytic activity. Some catalysts show better removal and photodegradation activity at high pH [50].

(iii) Concentration of photocatalyst: Photocatalytic efficiency can be controlled by photocatalyst concentration. The catalyst loading in a reaction controls the activity in heterogeneous photocatalysis. Initially, higher the concentration of photocatalyst, the higher will be the photodegradation. Later, the photodegradation rate decreases as the concentration of catalyst is increased, due to unfavorable scattering of photons, thereby reducing the penetration of light due to excessive loading of photocatalyst. Therefore, it is necessary to calculate the optimum concentration of photocatalyst for loading. The optimum concentration can be calculated by considering the light intensity, morphology of nanoparticles and design of biomaterial-assisted photocatalysis, as inconsistent loadings have been reported [41].

(iv) Duration and intensity of Light: The rate of reaction can be controlled by intensity of photons and absorption of these photons by the photocatalyst. At low light intensity, the generation of electron–hole pair becomes predominant; the rate of reaction increases linearly as light intensity is increased, thereby, formation of electron–hole pair tends to compete with its recombination and lowers the effect of intensity of light on rate of reaction; hence, the rate constant becomes inversely related to the intensity of light and time of irradiation. It occurs due to the pseudo-first order kinetics of photocatalysis. After repeated exposure of high intensity photons and deposition of byproduct, the photocatalysis deactivates [30].

(v) The impact of adding oxidizing species: The addition of oxidizing species to suspensions minimizes the recombination of electrons and holes. Mostly, hydrogen peroxide is used as an additive in catalytic reactions for production of hydroxyl radicals either through reduction on the surface by accepting electron and promoting charge separation in conduction band, or by decomposition. The degradation increases linearly with its concentration, until critical concentration is reached; the electron acceptors start scavenging radicals to reduce photocatalysis [6].

(vi) Temperature: Temperature of the reaction affects the rate of photocatalysis and follows first-order rate constant, increasing linearly as the temperature increases. At high temperature, the rate of reaction decreases, due to lower adsorption of molecules at the surface and loss of charge carriers on the surface. Therefore, specific temperature is favorable for photocatalysis. However, as the temperature lowers, activation energy is also increased [41].

(vii) Adsorption: By reducing the size of particle of photocatalyst, the rate of diffusion of photogenerated electron–hole pair can be controlled. The morphology of the particle, surface area, porosity and other reaction parameters (reaction temperature, pH of system and inorganic solvents) greatly impact photocatalytic reactions. Considering the morphology and properties of surface can be helpful in developing highly selective photocatalytic systems. Additionally, it would be economical and recyclable by using bioinspired materials as support of photocatalysts. Higher photocatalytic activity can be obtained by increasing surface area, porosity and adsorption capacity of the photocatalyst. Some supports have comparatively higher adsorption capacity, and thus, are brought to the surface for maximum photocatalysis, with more hydroxyl radical generation onto the surface and immediate reaction capability than others. Other than support, the thickness of the adsorbed layer also influences photocatalysis.

(viii) Dissolved oxygen: Dissolved oxygen helps scavenge electrons by trapping electrons from conduction band and helping reduce recombination of electron–hole pair. Thus, photodegradation increases when dissolved oxygen is increased in solution. The rate of transfer of electrons from the surface of electrons is increased as concentration of dissolved oxygen increases. Dissolved oxygen can stabilize the free radicals by biomineralization and direct photodegradation [29].

1.7 Challenges in enhancing photocatalytic properties

Different biomaterial-assisted photocatalysts have emerged as revolutionary catalysts, due to their properties like higher stability, huge bandgap, multiple functions and less toxicity. Nanomaterials have proved to be advantageous for photocatalysis, due to assorted morphology. Yet, the efficacy of photocatalyst is lower due to some associated challenges, such as mismatch between the bandgap and spectrum, insufficient separation of charges and instability of biomaterial (Figure 1.1).

1.7.1 Spectral mismatch

The bandgap energy regulates absorption of incident photons, and hence, to regulate the bandgap energy is a challenge for an efficient photocatalyst. The spectrum produces different wavelengths including IR (55%), visible (40%) and UV (5%) wavelength. Mostly, the photocatalysts are designed to have high energy bandgaps to absorb maximum UV wavelength, while the spectrum has more IR and visible rays. This phenomenon is defined as spectral mismatch. Therefore, remodulation of the bioinspired photocatalysts for spectral match is a fundamental challenge. Modulating the PCs so as to absorb maximum photons will be beneficial [51].

1.7.2 Recombination of the photogenerated charges

The recombination of electron–hole pair poses another challenge in modulation of PCs. Normally, the electron–hole pair is generated as the photons incident on the surface takes place; photogenerated electrons tend to migrate toward surface, and oxidation and reduction reactions take place. During this process, sometimes, photogenerated charge carriers may recombine either in bulk or on the surface. The energy is dissipated in the form of heat, suppressing the photocatalysis. In modulation of photocatalysts, there is no force to separate charge carriers. These charge carriers, however, are randomly involved in reaction and lead to nonproductive recombination, causing production of defected crystals. It is, therefore, necessary to reduce the chances of recombination to enhance the photocatalysis [24].

1.7.3 Low surface area

Surface area is another factor that influences photocatalysis activity by altering the interface chemistry, absorption of photons and rate of energy transfer. Therefore, to maintain higher surface area at the interface is another challenge in modulation of PCs, as all the reactions take place at the surface only. The surface area is directly proportional to photocatalytic activity [14].

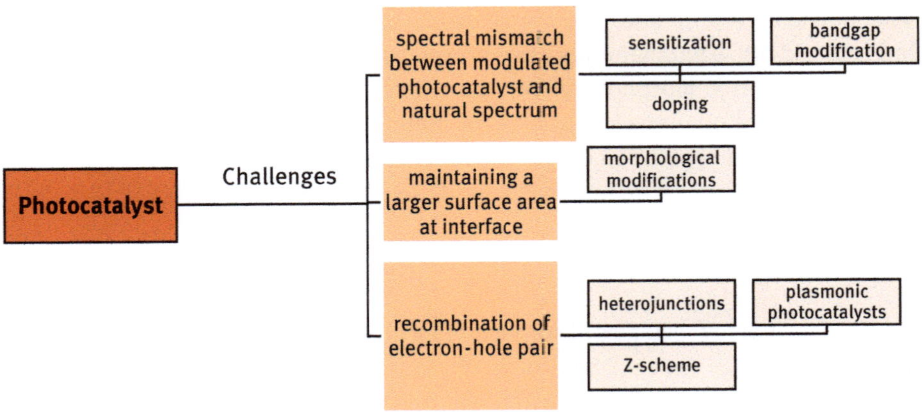

Figure 1.1: Challenges and strategies for an improved photocatalysis.

1.8 Techniques to increase photocatalytic efficiency

Different strategies have been proposed to enhance the biomaterial-assisted photocatalysis under UV or visible light so far, by tuning the bandgap according to natural spectra. It could be one of the most effective strategies based on remodulation of its structure; to achieve this, band structure can be modified to harvest maximum photons at the interface [4, 51–53].

1.8.1 Electronic band structure modification

Both conduction and valence band modulations can enhance the net photon harvesting efficiency. It can be achieved by adding specific impurity for induction of localized electronic state between ground state and excited state for electrons to reduce the overall bandgap. Consequently, it shifts the absorption from UV region to visible region and leads to better absorption. The level of doping and modifications of photon harvesting ability modifies the elemental properties by formation of homojunctions at the interface, by a well-controlled method [4].

1.8.2 Increasing charge separation

In photocatalysis, charge separation is a critical factor, having multiple phases and phase junctions. By increasing charge separation, the efficacy can be improved. It is a widely used approach for controlled photocatalysis [21]. For example, TiO_2, known for its use as photocatalyst, has three forms: brookite, rutile and anatase. Degussa P25 is the commercial replacement for TiO_2; it structurally resembles anatase–rutile mixture and is used for its high photocatalytic properties. Its performance can be related to the synergetic effect and its reduced bandgap. Fabrication can increase the overall activity by enhancing the charge separation [51].

1.8.3 Band energy engineering

This can enhance the photocatalysis by modifying the light absorption and redox potential. For effective photon harvesting, band energy engineering can be a powerful approach for developing photocatalysts of visible active region. To engineer the band energy, three approaches can be used: (i) nonmetal doping to modify its valence band; (ii) metal doping to modify conduction band; and (iii) continuous modification of valence band and conduction band [54].

1.8.4 Improved charge separation through heterogeneous networks

By fabricating this, the challenges in photocatalysis can be overcome. Heterojunctions are very beneficial in increasing the absorption by sensitization, enhancing optical properties by stabilizing the photocatalyst and charge separation, enhancing charge separation by junction with metals or semiconductors and improving overall stability by surface passivation to protect the interface from direct degradation and side reactions [41].

1.8.5 Morphology control

Lower the size of any particle, larger will be its surface area, and accordingly, greater will be its number of available active sites. Particle morphology is another strategy to cope up the challenges, hence the overall activity [21]. A suitable concentration gradient is to be maintained for effective charge separation on the surface; it is correlated with structure, particle size and surface properties. Mostly, researchers rely on the study of the morphologies to study the photocatalytic activity. Synthesis of modulated photocatalysts is another strategy to enhance light absorption and optimize charge separation by providing better porosity, larger surface area, and interconnections. Biotemplating is used to get morphology-controlled bioinspired materials with high structural and functional strength. The advantage of biotemplating is the result of final stage of photocatalyst, which can be tuned for its performance by selection of template with selective morphology.

1.9 Conclusion

Scientists are keen to develop bioinspired materials as a substitute for synthetic materials for different applications with a motive "return to nature." This chapter has elaborated different kinds of bioinspired materials: biopolymers, biochar, biomolecules, enzymes, peptides and proteins, for their role in photocatalysis. Such materials not only come from nature but also have a special interaction with biomolecules. Their physiochemical properties, surface area and structure provide insightful advantages in photocatalytic systems. The extensive research on bio-based photocatalysts and rapid advancement in their design and fabrication has enabled environmentalists to modulate the properties of biomaterials by preserving their activity and structure, not only during its synthesis but also during the light exposure. Subsequent sections of the chapter focused on various modes of synthesis with their advantages and disadvantages. The bio-based photocatalysts have been used for photocatalytic removal of contaminants with high performance. However, even after considerable advancement, environmentalists are still trying to develop strategies for different challenges in modulation of bioinspired materials to assemble biological and photocatalytic properties, without losing their individuality. Enhanced research is needed to enhance their photocatalytic potential as compared to more capable "star" photocatalysts. Their industrial feasibility is still under doubt, due to the difficulty in mass manufacturing, nature and activity. Thus, research is needed to elucidate the complex mechanism of immobilized biomolecules, specifically enzymes in photo degradation. This could be done by collaboration among chemists and scientists to develop novel bio-based, cost- effective and eco-friendly photocatalysts by developing biohybrid technologies.

References

[1] Ahmad, A., et al., *A novel study on synthesis of egg shell based activated carbon for degradation of methylene blue via photocatalysis*. Arabian Journal of Chemistry, 2020. **13**(12): p. 8717–8722.

[2] Hu, P., Hu, X., Chen, C., Hou, D. and Huang, Y., 2014. Biomaterial-assisted synthesis of AgCl@ Ag concave cubes with efficient visible-light-driven photocatalytic activity. *CrystEngComm*, **16**(4), pp.649–653.

[3] Kumar, A., Sharma, G., Naushad, M., Ala'a, H., García-Peñas, A., Mola, G.T., Si, C. and Stadler, F.J., 2020. Bio-inspired and biomaterials-based hybrid photocatalysts for environmental detoxification: A review. *Chemical Engineering Journal*, **382**, p.122937.

[4] Kumar, A., Sharma, G., Naushad, M., Ala'a, H., García-Peñas, A., Mola, G.T., Si, C. and Stadler, F.J., 2020. Bio-inspired and biomaterials-based hybrid photocatalysts for environmental detoxification: A review. *Chemical Engineering Journal*, **382**, p.122937.

[5] Hitam, C.N.C. and Jalil, A.A., 2022. Recent advances on nanocellulose biomaterials for environmental health photoremediation: An overview. *Environmental Research*, 204, p.111964.

[6] Patel, H., 2021. Blood biocompatibility enhancement of biomaterials by heparin immobilization: a review. *Blood Coagulation & Fibrinolysis*, **32**(4), pp.237–247.

[7] Tavolaro, P., Tavolaro, A. and Martino, G., 2009. Influence of zeolite PZC and pH on the immobilization of cytochrome c: A preliminary study regarding the preparation of new biomaterials. *Colloids and Surfaces B: Biointerfaces*, **70**(1), pp.98–107.

[8] Bibi, S., et al., *Synthesis of Cr doped LiMnPO4 cathode materials and investigation of their dielectric properties*. International Journal of Energy Research, 2021.

[9] Raees, A., et al., *Synthesis and characterization of Ceria incorporated Nickel oxide nanocomposite for promising degradation of methylene blue via photocatalysis*. International Journal of Environmental Science and Technology, 2021: p. 1–8.

[10] Beena, V., et al., *Photocatalytic dye degradation and biological activities of Cu-doped ZnSe nanoparticles and their insights*. Water, 2021. **13**(18): p. 2561.

[11] Sharma, S.K., Sehgal, N. and Kumar, A., 2003. Biomolecules for development of biosensors and their applications. *Current Applied Physics*, **3**(2–3), pp.307–316. Zhang, X.Z., et al., Synthesis, characterization and wettability of Cu-Sn alloy on the Siimplanted 6H-SiC. Coatings, 2020. **10**(9): p. 906.

[12] Reshmy, R., Philip, E., Madhavan, A., Sindhu, R., Pugazhendhi, A., Binod, P., Sirohi, R., Awasthi, M.K., Tarafdar, A. and Pandey, A., 2021. Advanced biomaterials for sustainable applications in the food industry: Updates and challenges. *Environmental Pollution*, **283**, p.117071.

[13] Javed, M.S., et al., *2D V2O5 ultrathin nanoflakes as a binder-free electrode material for high-performance pseudocapacitor*. Ceramics International, 2021.

[14] Mukhopadhyay, S., Gowtham, S., Scheicher, R.H., Pandey, R. and Karna, S.P., 2010. Theoretical study of physisorption of nucleobases on boron nitride nanotubes: a new class of hybrid nano-biomaterials. *Nanotechnology*, **21**(16), p.165703.

[15] Lee, H.P. and Gaharwar, A.K., 2020. Light-responsive inorganic biomaterials for biomedical applications. *Advanced science*, **7**(17), p.2000863.

[16] Syah, R., et al., *Incorporation of Bi2O3 residuals with metallic bi as high performance electrocatalyst toward hydrogen evolution reaction*. Catalysts, 2021. **11**(9): p. 1099.

[17] Shi, Z.Q., Li, H., Xue, C.L., Yuan, Q.Q., Lv, Y.Y., Xu, Y.J., Jia, Z.Y., Gao, L., Chen, Y., Zhu, W. and Li, S.C., 2020. Tuning the electronic structure of an α-antimonene monolayer through interface engineering. Nano Letters, **20**(11), pp.8408–8414.

[18] Javed, M.S., et al., *Design and fabrication of bimetallic oxide nanonest-like structure/carbon cloth composite electrode for supercapacitors*. Ceramics International, 2021. **47**(21): p. 30747–30755.
[19] Bibi, S., et al., *Photocatalytic degradation of malachite green and methylene blue over reduced graphene oxide (rGO) based metal oxides (rGO-Fe3O4/TiO2) nanocomposite under UV-visible light irradiation*. Journal of Environmental Chemical Engineering, 2021. **9**(4): p. 105580.
[20] Ahmad, I., et al., *Lanthanum-zinc binary oxide nanocomposite with promising heterogeneous catalysis performance for the active conversion of 4-nitrophenol into 4-aminophenol*. Coatings, 2021. **11**(5): p. 537.
[21] Tung, W.S., W.A. Daoud, and S.K. Leung, *Understanding photocatalytic behavior on biomaterials: Insights from TiO2 concentration*. Journal of Colloid and Interface Science, 2009. **339**(2): p. 424–433.
[22] Kumar, A., et al., *Bio-inspired and biomaterials-based hybrid photocatalysts for environmental detoxification: A review*. Chemical Engineering Journal, 2020. **382**: p. 122937.
[23] Zhao, Y., et al., *Fabrication of biomaterial/TiO2 composite photocatalysts for the selective removal of trace environmental pollutants*. Chinese Journal of Chemical Engineering, 2019. **27**(6): p. 1416–1428.
[24] Cai, Y., *Titanium dioxide photocatalysis in biomaterials applications*. 2013, Acta Universitatis Upsaliensis.
[25] Zhu, X.-D., et al., *Photocatalytic degradation of tetracycline in aqueous solution by nanosized TiO2*. Chemosphere, 2013. **92**(8): p. 925–932.
[26] Mian, M.M. and G. Liu, *Recent progress in biochar-supported photocatalysts: synthesis, role of biochar, and applications*. RSC Advances, 2018. **8**(26): p. 14237–14248.
[27] Liu, Z., W.R. Leow, and X. Chen, *Bio-inspired plasmonic photocatalysts*. Small Methods, 2019. **3**(3): p. 1800295.
[28] Zhang, J., F. Chen, and B. He, *Photocatalysis*. 2018: Springer.
[29] Ameta, R., et al., *Photocatalysis*, in *Advanced oxidation processes for waste water treatment*. 2018, Elsevier. p. 135–175.
[30] Di Paola, A., et al., *Heterogeneous photocatalytic degradation of nitrophenols*. Journal of Photochemistry and Photobiology A: Chemistry, 2003. **155**(1–3): p. 207–214.
[31] Zhao, J., C. Chen, and W. Ma, *Photocatalytic degradation of organic pollutants under visible light irradiation*. Topics in Catalysis, 2005. **35**(3–4): p. 269–278.
[32] Kosera, V.S., et al., *Triclosan degradation by heterogeneous photocatalysis using ZnO immobilized in biopolymer as catalyst*. Journal of Photochemistry and Photobiology A: Chemistry, 2017. **344**: p. 184–191.
[33] Primo, A., et al., *Efficient visible-light photocatalytic water splitting by minute amounts of gold supported on nanoparticulate CeO2 obtained by a biopolymer templating method*. Journal of the American Chemical Society, 2011. **133**(18): p. 6930–6933.
[34] Dong, W., et al., *A performance study of enhanced visible-light-driven photocatalysis and magnetical protein separation of multifunctional yolk–shell nanostructures*. Journal of Materials Chemistry A, 2013. **1**(34): p. 10030–10036.
[35] Ding, X., et al., *Concurrent asymmetric reactions combining photocatalysis and enzyme catalysis: direct enantioselective synthesis of 2, 2-disubstituted indol-3-ones from 2-arylindoles*. Angewandte Chemie, 2019. **131**(1): p. 124–130.
[36] Ariza-Tarazona, M.C., et al., *New strategy for microplastic degradation: Green photocatalysis using a protein-based porous N-TiO2 semiconductor*. Ceramics International, 2019. **45**(7): p. 9618–9624.

[37] Jin, B., et al., *Physicochemical stability and antioxidant activity of soy protein/pectin/tea polyphenol ternary nanoparticles obtained by photocatalysis*. International Journal of Biological Macromolecules, 2018. **116**: p. 1–7.

[38] Carré, G., et al., *TiO2 photocatalysis damages lipids and proteins in Escherichia coli*. Applied and Environmental Microbiology, 2014. **80**(8): p. 2573–2581.

[39] Kadam, A.N., et al., *Biogenic synthesis of mesoporous N–S–C tri-doped TiO2 photocatalyst via ultrasonic-assisted derivatization of biotemplate from expired egg white protein*. Applied Surface Science, 2020. **518**: p. 146194.

[40] Wang, Y., et al., *Biomolecule-controlled hydrothermal synthesis of C–N–S-tridoped TiO2 nanocrystalline photocatalysts for NO removal under simulated solar light irradiation*. Journal of Hazardous Materials, 2009. **169**(1–3): p. 77–87.

[41] Kim, J.R. and E. Kan, *Heterogeneous photocatalytic degradation of sulfamethoxazole in water using a biochar-supported TiO2 photocatalyst*. Journal of Environmental Management, 2016. **180**: p. 94–101.

[42] Manimohan, M., et al., *Synthesis and characterisation of novel Cu (II)-anchored biopolymer complexes as reusable materials for the photocatalytic degradation of methylene blue*. RSC Advances, 2020. **10**(31): p. 18259–18279.

[43] Xiao, Q., et al., *Solar photocatalytic degradation of methylene blue in carbon-doped TiO2 nanoparticles suspension*. Solar Energy, 2008. **82**(8): p. 706–713.

[44] Othman, S.H., et al., *Dispersion and stabilization of photocatalytic TiO2 nanoparticles in aqueous suspension for coatings applications*. Journal of Nanomaterials, 2012. **2012**.

[45] Di, J., et al., *The synergistic role of carbon quantum dots for the improved photocatalytic performance of Bi 2 MoO 6*. Nanoscale, 2015. **7**(26): p. 11433–11443.

[46] Abdel-Wahab, M.S., et al., *Enhanced the photocatalytic activity of Ni-doped ZnO thin films: Morphological, optical and XPS analysis*. Superlattices and Microstructures, 2016. **94**: p. 108–118.

[47] Gnaser, H., et al., *Photocatalytic degradation of methylene blue on nanocrystalline TiO2: surface mass spectrometry of reaction intermediates*. International Journal of Mass Spectrometry, 2005. **245**(1–3): p. 61–67.

[48] Moctezuma, E., et al., *Photocatalytic degradation of paracetamol: intermediates and total reaction mechanism*. Journal of Hazardous Materials, 2012. **243**: p. 130–138.

[49] Li, Y., et al., *Photocatalytic hydrogen generation in the presence of chloroacetic acids over Pt/TiO2*. Chemosphere, 2006. **63**(8): p. 1312–1318.

[50] Chen, S. and Y. Liu, *Study on the photocatalytic degradation of glyphosate by TiO2 photocatalyst*. Chemosphere, 2007. **67**(5): p. 1010–1017.

[51] Hoven, V.P., et al., *Surface-charged chitosan: Preparation and protein adsorption* Carbohydrate Polymers, 2007. **68**(1): p. 44–53.

[52] Daud, M.K., et al., *Drinking Water Quality Status and Contamination in Pakistan*. BioMed Research International, 2017. **2017**: p. 7908183.

[53] Du, J., et al., *Hierarchically ordered macro- mesoporous TiO2- graphene composite films: improved mass transfer, reduced charge recombination, and their enhanced photocatalytic activities*. ACS Nano, 2011. **5**(1): p. 590–596.

[54] Wang, H., et al., *Semiconductor heterojunction photocatalysts: design, construction, and photocatalytic performances*. Chemical Society Reviews, 2014. **43**(15): p. 5234–5244.

Sadaf Tariq, Awais Ahmad, Rafael Luque, Mabkhoot Alsaiari,
Farid A. Harraz

2 Adsorption study via biomaterials

Abstract: With advancements in Nanochemistry, formulations of biomaterials have gained importance in biomedical applications for therapeutic and nontherapeutic purposes. Biomaterials have been studied for their regenerative applications in tissue engineering to provide scaffold and support for diverse cellular activities, as a biotherapeutic agent. After implantation, different biomolecules are adsorbed onto the surface of biomaterial for different activities. Irrespective of activity, biomolecules are either physically or chemically adsorbed on the surface. Physical adsorption results from entrapment of the biomolecule or forming layers, while chemical adsorption involves a chemical reaction for tethering biomolecule to surface. The adsorption and consequent activity of biomaterial are influenced by the surface chemistry and biocompatibility of biomaterial surface and biomolecules in order to avoid nonspecific adsorption of biomolecules. Thus, surface chemistry characterizes encouraging directions to specifically engineer its biofunctionality at the biomaterial–cellular interface to control related biological response, without changing overall properties of biomaterial. The concept of adsorption methods as also surface modifications to avoid nonspecific adsorption and the techniques to study the mechanism of adsorption and chemistry of adsorbed and adsorbing molecules are provided in this chapter.

Keywords: Biomolecules, biomaterials, nanochemistry, adsorption, surface chemistry, biocompatibility

2.1 Introduction

Biomaterials are nonviable substances that interact with the biological systems, developed for different medical, nonmedical and pharmaceutical purposes, in both therapeutic and nontherapeutic areas [1–3]. Biomaterials are natural or non-natural materials used to replace or repair damaged body tissues and organs and can enhance or replace certain reduced or lost functions [4, 5]. Living cells are easily combined with the biomaterials, particularly when their surfaces comprise cell adhesion fragments, proteins or peptides [1, 6]. Different biological fields such as cell culture and tissue engineering include various examples of different biological cells–surface interactions [7, 8]. This hybrid combination of natural cells with synthetic materials confers the biological functionality of cells to synthetic biomaterials [9–11]. Thus, by understanding the biomaterial adsorption, biomaterial–biomolecule interaction [12] and surface chemistries, biological mechanisms occurring between biomaterials and

https://doi.org/10.1515/9783110768749-002

biomolecules [13] and molecular events can be the central key to the successful practice of biomaterials in clinics [14–16].

Surface functionalization enables the interaction of biomolecules with biomaterials in different orders of magnitude [17, 18]. Within a few seconds of implantation of any biomaterial, an abundance of serum proteins is adsorbed on the surface of biomaterial instantly, leading to a series of ensuing effects such as cellular adhesions, coagulation activity [19, 20], complement system activation, leukocytes reactions, Intrinsic coagulation, fibrinolytic activity and platelets activation [10, 15, 21, 22]. The composition of adsorbed proteins, consequently, determines the response of the cells lying under the biomaterial scaffold, forming either effective tissue integration or disruptive body reactions. Unfortunately, if the nonspecific adsorption occurs, an associated inflammatory response happens [23]; it can be minimalized by modifications of the biomaterial surface chemistry. Furthermore, biomaterial surfaces are functionalized in a way to present them to a specific protein or peptide sequence, which encourages a particular cell adhesion [1, 24, 25]. In an effort to control the biological–material interface in a better way, different physical and chemical surface biocompatibility strategies are developed [26, 27]. Promoting specific adsorption, controlling cellular behavior and avoiding any adverse host responses are considered key issues during the design of any biomaterial for biological applications [6, 28]. Though other properties such as matrix composition and its mechanical model play crucial roles in determining the biomaterial's overall biocompatibility with the surface and its success towards engineered tissue scaffold, properties of adsorptive surfaces are one of the most critical factors on how any biomaterial tends to interact with its surrounding environment [9, 29, 30].

This chapter will focus on the adsorption study involving physical adsorption via surface adsorption, physical entrapment, self-assembled monolayer and layer-by-layer deposition and chemical adsorption of biomolecules on biomaterial surfaces exclusively, to create an understanding about the biocompatibility of surface and biomolecule, along with specific surface modifications for effective adsorption of biomaterials for different applications requiring the bioactivity of the adsorbed biomolecules. This chapter will exclusively elaborate the techniques to monitor the mechanism of adsorption and the molecules adsorbing on biomaterial surface or already adsorbed molecules on the surface.

2.2 Fundamentals of adsorption

To begin with, adsorption of a biomolecule to a biological material surface is mainly affected by the bulk protein content. However, this process is much more intricate when structurally and functionally different proteins are present because, in this case, many proteins get involved in the reaction [31]. Another significant factor in protein–biomaterial interaction is the diffusion and effusion rates of the protein in cell – these rates are controlled primarily by the protein itself [32]. Smaller proteins easily get to the surface of the cell as these are faster and adsorbed onto the surface compared to bulkier proteins. Again, another crucial factor, and definitely the most important for long-term concerns, is the affinity of a specific protein for biomaterial surface [33]. The affinity of any protein for biomaterial surface defines how probable that protein's adsorption is and its strength of adherence to any biomaterial surface [34]. The proteins with higher affinity for biomaterial are easily adsorbed and stay on the biomaterial surface. This interaction varies immensely, based on the biocompatibility of protein and biomaterial surface [33, 35].

Normally, after implantation of biomaterial in biological system, the biomaterials tend to adsorb proteins. Many proteins may adsorb on the surface of biomaterials nonspecifically to denature the biomaterial. For example, immune proteins like platelets and monocytes tend to adsorb on biomaterial, which concurrently causes an elevation in cytokines and other inflammatory cells and proteins in the human body. The implanted biomaterial can cause either specific or nonspecific response in the body, depending on the reaction between protein and surface of biomolecules [1, 36]. Mostly, the implanted biomaterials are designed in a way that avoids their phagocytosis by macrophages, so their size is kept bigger than the macrophages. The cytokines are accumulated at the surface of biomaterials at the site permanently and use collagenous fibrous system for cellular adhesions [7].

Proteins adsorbed to biomaterial form bonds; high affinity proteins form stronger bonds while low affinity proteins tend to make weaker bonds. Protein level, rates of diffusion and effusion and protein affinity – these aspects together confer that proteins adsorbed on biomaterial surface can replace each other, over time (Vroman effect) [32, 37]. The interactions formed between a protein and biomaterial are established throughout the adsorption mechanism, are not stable, can be broken and reformed randomly over time [35, 38]. However, any protein is unlikely to desorb off surface, because all connections between both the surface and the protein can be broken down at the same time. Contrary to other proteins, new bonds may be formed at available sites [34]. If the novel protein demonstrates greater affinity for the accessible site, the protein will take over the available site even at lower affinity [38]. Nonetheless, for in vivo system and implantation, the adsorbed layer of biomolecules on the surface of biomaterials is developed specifically in the form of small size fragments, sensors and kidney dialyzers.

The biological system and its environment affect the process of adsorption, such as conformation, stability and interactions, to a great extent. Temperature has an important effect on adsorption; high temperature can cause denaturation, and reduced bioactivity causes ineffective results. Different biomolecules are affected at different rates at varied temperature and pH; the environmental effects cause denaturation, aggregation and total destruction of conformation of proteins. The biomaterials are designed to carry out their bioactivity in a range of temperature and pH. The change in pH causes variations in interaction in cell–biomaterial interface. Adsorption of any biomolecule would be highest at a specific temperature and pH. Last but not the least, the surfactants can also cause alterations in rate of adsorption to biomaterial by competing with other biomolecules for available sites or by directly binding to biomolecules [39, 40].

2.3 Adsorption of biomolecules on surface of biomaterials

The adsorption of biomolecules occurs by hydrophobic, ionic or covalent interactions between biomolecules and the surface of biomaterials where the binding takes place at the activated functional groups [6]. Noncovalent interactions occur, leading to passive adsorption [29]. Consequently, the covalent interactions are highly stable and preferable, as the chances of nonspecific adsorption are very low. The adsorption events manifest a number of characteristics – the required reaction should occur quickly and require comparatively lower amount of reagents [41]. The process of adsorption occurs easily, when common functional groups such as amines, carboxylic acids, thiols and alcohols are found. Occasionally, the physical adsorption involves different molecules of stronger affinity, verging on different covalent interaction levels, as in biotin–streptavidin interactions. Thus, the "immobilization" term can refer to either short-term or any permanent localization of the biomolecules, with or without support [41, 42]. In a drug delivery system, the immobilized drugs are only supposed to be released, either over a short term or may be for longer duration, while any immobilized enzyme, peptide or protein carrying the ability of cell adhesion in any organ is supposed to be designed to attach or entrap, within a specified duration [41]. Both physical and chemical immobilizations lead to comparatively long-term or permanent retention time, especially if immobilized biomolecules are large [42, 43].

2.3.1 Physical adsorption

Physical adsorption occurs through van der Waal's interactions, electrostatic interactions and affinity recognition. Physical adsorption is simpler than chemical adsorption and less likely to affect related biomaterial properties. A thin film of coatings is formed on the surface of biomaterial in physical adsorption, instead of any alteration in atoms and molecules present on the surface. These films and coatings are formed as a result of electrostatic, hydrogen interactions and nonchemical strategies and occur during physical entrapment and layer assembly. For physical immobilization, the biomaterials are exposed to air to increase their hydrophilicity

[41, 43]. Noncovalent adsorption is based on electrostatic interactions between surface and biomolecules. For example, gelatin microparticles are loaded with transforming growth factor-1 (TGF-1) and encapsulated in polyethylene glycol (PEG) hydrogels at a pH of 7.4 to form an ionic complex. The interactions are developed due to negative charge of the functional groups of gelatin and TGF-1 positive charge [41, 44].

2.3.1.1 Physical entrapment

The physical enclosure of biomolecules on the surface of biomaterial in a small space, usually at the matrix and membrane, is a major method of physical entrapment. Physical entrapment forms a comparatively stable and noncovalent interaction with the surface of various biomaterials and requires cautious selection of active functional groups to solubilize the biomolecules integrated on the surface, preventing its degradation [45]. The biomaterial is swelled in a suitable solution, a nonsolvent is added, resulting in shrinkage of biomaterial, and the biomolecules are entrapped on surface. This technique has been effectively used to integrate biomolecules (chitosan, gelatin and heparin) on surfaces of biomolecules, resulting in cellular adhesion [46]. Although physical entrapment forms relatively more stable surface interactions than passive adsorption, it is unable to work properly for many biomaterials and other surface-functionalized molecules and allows the biomolecules to pass through it easily, but retains the biomolecules in a confined network too. Hence, it works as a restriction for biomolecules movement. Characteristically, it can improve the stability and reduce the denaturation of biomaterial. Thus, by encapsulating the biomaterial, it becomes more fruitful [46, 47].

It is an irreversible method of immobilization of biomolecules, where entrapping takes place on or in the surface of biomaterials, either on the membrane or inside the lattice. Physical entrapment is followed by a number of biomaterials including biopolymers, sol–gels, biocomposites [48] and inorganic biomaterials. Proper adsorption can be done when the pore size of biomaterial is kept small. Other modifications can enhance or decrease the effectiveness [49].

2.3.1.2 Surface adsorption

Physical adsorption of biomolecule on the surface is one of the simplest methods of adsorption. Simply, a coating of biomolecules is formed on the biomaterial surface, aided by some proteins such as fibrinogen, fibronectin, laminin and collagen for improved cellular adhesions [50]. As already described, the adsorption of any biomolecule to a surface is controlled by different factors such as surface properties, properties of biomolecules, coupling method and environmental conditions. The interplay of multiple factors tends to make it problematic to control the adsorption, resulting in low coating efficiency. More essentially, passive adsorption is comparatively unstable and desorbs over time [41, 51].

2.3.1.3 Self-assembled monolayers

Self-assembled monolayers are spontaneously formed on the surface of biomaterials as highly ordered, thin surface coatings. Mainly, the biomolecules comprise of a certain structure having a chemical head functional group, which tends to bind with the biomolecule with different strengths – strong to moderate [52]. The tail group possesses a terminal functional group that allows increased strength and stability [53]. The chemical groups involved in forming self-assembled monolayers can be precisely selected, by taking alkane thiols and disulfides on metal surfaces, for example. Though self-assembled monolayers with surface functionalization of many biomaterials have limited applications, novel classes of amphiphilic biopolymers, which can form similar layers on biopolymeric surfaces with hydrophobic interactions [41, 54] are being developed.

2.3.1.4 Electrostatic layer-by-layer deposition

Electrostatic layer-by-layer deposition involves the consecutive adsorption on the surface of biomaterial by forming alternative layers of opposite charges containing polyelectrolytes [55]. Temporarily, charged biomaterial is submerged in a polyionic solution, adsorbed on the surface, yields the first layer by reversing the net charge and unbound electrolytes are rinsed away [56]. A second layer is formed by immersing it in another polyionic solution containing opposite charges. If the first layer was formed by submerging it in polycationic solution, then the second layer is supposed to be formed by submerging it in polyanionic solution [57]. This cycle of layer formation is repeated several times, until the desired form is obtained. The properties of layers can also be fine-tuned by the photoelectrolytes. The layers can be produced by grafting, spraying and coating. Different factors like surface chemistry, permeability of membranes and thickness can be tailored too [52]. Multilayers can be formed and modified for a variety of purposes, by having the conformation of biomaterial and biological system in view [56]. More attractively, many coatings like albumin coating, chitosan coating and phospholipidic coatings have been designed by using electrostatic interactions, hydrogen bonding and hydrophobic interactions [41, 55].

2.3.2 Chemical adsorption

Chemical adsorption takes place via comparatively weaker interactions than physical adsorption and covalent binding with functional groups (Table 2.1). The process, sometimes, tends to form permanent immobilization with specified conformation [58]. Chemical adsorption is the most commonly used method, as it works as powerful link and prevents the release of biomolecules before reaction, once attached. Once the reaction completed, stability of the biomolecule is enhanced by multiple covalent bonds. It can be understood well by taking silica and chitosan coatings as examples. The biomolecules are localized on the surface of biomaterials, and it enhances further attachment and support [42].

After covalent bonding, the bioactivity of biomaterial depends on the basic conformation of biomaterial, the coupling method of biomolecules, composition of the biomaterial, specific environmental conditions during coupling and available functional groups on the surface [59]. The available reactive functional groups provide room to interact without modifications, or the matrix is modified in a way that generates functional groups [58]. The electrophilic groups on the biomaterial surface tend to react with nucleophiles on the biomolecule and vice versa [42, 60].

Table 2.1: An outline of comparison of chemical surface functionalization with respective advantages.

Chemical adsorption	Hydrolysis	Activates surface for adsorption and grafting
	Oxidation	Activates functional groups to enhance adsorption
	Grafting-to	Creates flexible antifouling and active surface to enhance adsorption
	Plasma treatment	Increases cell adhesion
	Aminolysis	Activated surfaces
		Antifouling and biomimetic coatings
	Grafting from	Flexible coatings
		High grafting densities

2.4 Overview of biomaterial surfaces related to adsorption: strategies, challenges and approaches

Whether the target is to prevent nonspecific adsorption of biomolecules, enhance only specific cellular adhesion, trigger specific biological reaction or allow delivery of any bioactive agent, the surfaces play a pivotal role. In order to increase the in vivo activity of surface modified, nonfouling, biocompatible, biomimetic and bioactive biomaterials, chemical and physical methods have been developed for surface modifications. Therefore, a brief note on surfaces of biomaterials, biocompatibility with the biomolecules, challenges and approaches is given in this chapter.

2.4.1 Biomimetic surfaces

Biomaterial surfaces can be modified to promote specific cellular adhesion and function by modifying their molecular sequences. It not only improves the chance of effective integration but also simulates the physiological interactions, as cell–matrix cellular adhesion. Surfaces with modified proteins or peptides sequences that mimic the matrix proteins or peptide sequences of biological system showed functionalized and improved cellular adhesion and viability for implantation and activity of biomaterials [61]. Coating the surfaces with some proteins commonly found in extracellular matrix such as fibronectin, collagen, and elastin has been an interesting approach to modify the surfaces into biomimetic surfaces [62]. These proteins are comprised of specific motifs involved in binding integrins with receptors to promote cellular adhesion. These surfaces can be developed and functionalized by surface adsorption, surface grafting and chemical conjugation. These biomimetic surfaces enhance cellular adhesions [61, 63].

2.4.2 Surface properties

The properties of biomolecules are as important as the properties of surfaces of biomaterials in adsorption. Both factors are determinants of the required function in the biological systems that preferentially adsorb [37]. It is to be noted that, after implantation, serum proteins are adsorbed on the biomaterial surface, and the water molecules in the biological system tend to interact with biomaterial. The hydrophobic biomaterials create a film or coating of water molecules around; water molecules interact both itself and with the surface. A film of water molecules is formed in an ordered form and low entropy. When the entropy is increased, so does the adsorption. Thus, the entropy acts as a limiting factor in adsorption of proteins to biomaterial hydrophobic surfaces. Accordingly, there is an elevated adsorption, and changes in conformation of biomaterial have been observed on these surfaces [64]. Conformational changes affect the activity of proteins and, thereby, adsorption is also affected, as the native state of protein requires being interacted [65].

In the case of hydrophilic surfaces, water molecules tend to make hydrogen bonding, and a competition arises among the water molecules or other biomolecules, causing a reduction in adsorption by specific biomolecules and changes in its conformation [66]. Furthermore, the surface of biomaterials has a specific charge on it due to specific active functional groups. As same charges repel each other, only the opposite charges help in adsorption and different interactions such as ionic, van der Waal's and hydrophobic interactions can occur [67, 68]. Meanwhile, negatively charged blood serum proteins, if there is a net negative charge on biomaterial surface, will cause reduction in adsorption of serum proteins; it allows the binding of other positively charged biomolecules to the surface of biomaterial. It is, however, a problem because protein containing positively charged domains amino acids may still be able to make some interactions [66, 69].

2.4.2.1 Effect of functional groups of biomaterial surface on adsorption

The functional groups on a biomaterial surface potentially not only affect adsorption of biomolecules specifically proteins, but also trigger the successive cellular adhesions. Surface chemistry, topography, conformation and bioactivity of the adsorbed biomolecules depend on the available functional groups on the surface. There is no direct relationship between protein concentration and the bioactivity at the surface – high concentration of proteins or other bioactive molecules at the site never confers high bioactivity [70, 71]. It could be a totally different story. In this chapter, different functional groups and their effects on the adsorption are discussed in following paragraphs.

Methyl, a hydrophobic functional group, binds fibrinogen and IgG and encourages higher production of inflammatory cells at the infection site, caused by its higher adsorption. It promotes protein adsorption for required cellular adhesions and increases accumulation of fibrinogen and platelets. Albumin, immunoglobulins and fibrinogen easily interact with methyl groups and generate inflammatory cells [72]. Hydroxyl, a hydrophilic functional group, reduces the activity of certain plasma proteins at the infection site, modifies conformation of fibronectin, and thereby, exposes the head and tail domains for cellular adhesions to strengthen and elevates the level of structural signaling components [73]. It also aids in better differentiation of osteoblasts and mineralization as compared to other functional groups. It causes reduction in levels of adsorption of plasma proteins, thereby leading to higher platelet adhesions. Amine, a hydrophilic group, binds to the positively charged proteins in blood and serum. It also binds strongly to the fibronectin to cause responses in immune cells by triggering inflammatory cells. Carboxyl, a hydrophilic group, binds to blood and serum protein, reacting especially more easily with fibronectin and albumin to elute them from the surface. The surfaces containing carboxyl terminals show enhanced cell growth, but the abundance of these terminals can halt the cellular growth by giving the membrane overall negative charge, inhibiting differentiation of osteoblasts and causes biomineralization. In the case of both hydrophobic and hydrophilic functional groups, a competition arises, as the concentration of hydrophilic groups decreases the protein adsorption, platelet activation and cellular adhesions. The presence of other biomolecules like water in the biological system also affects the rate of adsorption, to a great extent [72, 74, 75].

2.4.2.2 Surfaces that inhibit nonspecific interactions

Surface modifications can prevent nonspecific interactions between biomolecules and biomaterials, helping improve their physiochemical interactions at molecular and cellular levels for better cellular adhesions and to avail many other properties; for example, it prevents corrosion of the surface and enhances surface functionalization by promoting the biocompatibility of biomaterial with biomolecule [76]. Different approaches have been consistently used to inhibit nonspecific adsorption on the surface of biomaterials by using phospholipids, hydrogels, polyethylene oxide (PEO), fluoropolymers and silicones [75, 76]. Some inorganic surfaces that resist uptake of proteins and biomolecules by cells highly, have been prepared through radio frequency plasma deposition of PEO films [77]. This approach has been used for developing thin and tightly adherent films coated on any material surface in a single step through dry process. This approach of RF-plasma immobilization has been described elsewhere [78]. Nevertheless, the biomaterial surfaces are coated, pressurized under a specific pressure, contaminants are removed, pumped under low pressure, another coating is deposited, evacuated and all the coated samples are removed. Such surface modifications resist interactions of proteins, platelets and other biomolecules. Surface functionalization studies revealed that layers similar to PEO films added on the biomaterial surfaces have substantial structural and functional similarity to PEO for inhibiting its recognition and avoiding any nonspecific interaction [78–80].

2.4.3 Nonfouling surfaces

Though hydrophilic surfaces are considered for their decreased adsorption, they are found to stimulate adsorption. The process of decreased adsorption by hydrophilic surfaces is known as nonfouling. This process can be well illustrated by considering proteins as adsorbed biomolecule [78]. As the serum proteins are adsorbed on cardiovascular biomaterials for immunological responses like blood clotting, if the strategy is to use such biomaterials inhibiting adsorption of serum proteins, PEG or PEO can be used. PEG is a hydrophilic biomolecule, used in drug delivery systems, as a surface in biomaterials and for various medical applications for its low protein adsorption [81]. PEG has low affinity for proteins due to its hydrophilicity. Once surrounded by water, these hydrophilic polymers are coated with water molecules and get super hydrated. When any protein comes in direct contact with a PEG surface, it requires energy to adsorb the protein, as it is thermodynamically unfavorable. But, with greater exposure time, some protein adsorbs to a smaller degree, such as tetraglyme (tetraethylene glycol dimethyl ether), known to produced cross-linked nonfouling assemblies, highly resists adsorption on in vitro surfaces and activates inflammatory responses by promoting the uptake of leukocytes. Though tetraglyme is highly resistant to blood serum proteins, the complementary system of the body adsorbs on tetraglyme surface, in substantial amounts. Monocytes cooperate with the complement factor, C3, in blood and serum for cellular adhesions, proposing that the binding of monocytes to adsorbed C3 on tetraglyme is the sole way to bind monocytes with tetraglyme surface, and is otherwise, not possible. The removal of tetraglyme surface from the biomaterial results in continuous loss of nonfouling activity. It suggests that PEG-coated surfaces must have gone through a significant C3 adsorption in vivo, to induce activation and adhesion of monocytes [82]. Therefore, modifying the biomaterial surfaces can reduce the chances of nonspecific protein adsorption and is done to avoid any adverse body response [80]. PEG reduces the interactions of nonspecific proteins on the surface coated by it and allows cell adhesion, while its nontoxic nature makes it favorable for use in biological systems. PEG and derivatives develop an associated hydrated layer and flexible polymer chains to resist proteins and cause steric repulsions. Different techniques can be used to make nonfouling surfaces coated with PEG, such as surface grafting, plasma spraying, physical adsorption, chemical adsorption, coating, self-assembled monolayers and polyelectrolyte multilayers [80]. PEG-based nonfouling coating lacks long-term stability, is degraded by oxidative and thermal damage and hydrolyzed easily. The basic problem with PEG chains is their autooxidation in aqueous solutions and reduction in protein resistance [83].

Zwitterions contain neutral charged units and a stable mixture of charged groups that interact by hydrogen and electrostatic interactions, strongly bound water layer and are important in forming nonfouling layer on the surface of biomaterials. The phosphorylcholine-containing biopolymers are the most studied nonfouling zwitterions, for their biomimetic surface [84]. Zwitterions containing biopolymers are useful for their nonfouling properties and are extensively used for surface modifications, as they resist protein adhesions with better potential as compared to PEG coatings and show a better stability, in vivo. These zwitterion biopolymers surfaces are developed on biomaterial surfaces by grafting, self-assembled monolayers, layer-by-layer films production and other depositions [85].

Polysaccharides have urged scientists to extensively investigate them for their nonspecific interactions, cell adhesions and nonfouling properties. Many polysaccharides and coated polymers have been created for their protein-repellent properties; they are coated on the surface of biomaterials [86]. The most commonly used nonfouling polysaccharides are dextran and its derivatives. Different methods have been employed to establish useful and effective polysaccharide-modified biopolymers coatings on the biomaterial surfaces, such as surface grafting, self-assembled monolayers, surface adsorption, grafted biopolymers and electrostatic layer-by-layer films [87]. Dextran and its derivatives are more stable, and after coating, reduce the cellular adhesions by proteins. Apart from the biomimetic nature of these biomaterials, nonfouling films that are formed by polysaccharides yield various advantages along with resistance to degradation by thermal changes and oxidation and long-term stability [88].

2.5 Molecular events during adsorption: biological and conformational changes

Molecules adsorbed to surfaces can go through biological and conformational changes because of different properties of adsorbed molecules: their structural high flexibility, low stability, tendency to form new bonds and variable thermodynamics. Biological changes are detected by detecting the changes in conformation of adsorbed molecules. Some molecules are easily adsorbed than others due to their low thermodynamic stability as soft proteins, while other molecules, due to their high stability, cannot be more persistently adsorbed [89]. By comparing the adsorptive nature of different biomolecules, it can be deduced that highly stable molecules are less adsorptive [90]. Hard proteins are highly stable even at different temperatures; do not unfold them at elevated temperature to prevent the denaturation. The understanding of surface chemistries of developed biomaterials according to the biocompatibility with the adsorbed molecule has been a major breakthrough in developing nonfouling biomaterial surfaces, in recent years. The adhesion of platelets requires adsorbed proteins, and therefore, protein repelling biomaterial is a topic of keen research [91]. Different compounds are studied and focused due to their protein-repellent capability and reduced platelets adhesion – polyethylene oxide (PEO) and different types of coatings of PEO on different surfaces (PEO-coated polylactide biopolymers, PEO-grafted polyurethanes or PEO-grafted polystyrene) or PEG, both in vivo and in vitro. Another major issue is faced in designing a biomaterial with the least affinity to fibrinogen, as fibrinogen tends to bind even at lowest amounts; the least binding can result in maximum platelets adhesion, even exposed in undiluted plasma [92]. Another point of interest is that PEO and PEG are used for short-term applications, as they are hydrolyzed and their protein repellency is lost, over time. Surfaces of polyurethane coated with PEO were found to show maximum protein repellency by adsorbing lipoproteins [very low density lipoproteins (VLDL), low-density lipoprotein (LDL), high-density lipoprotein (HDL)]; an unusual high concentration of lipoproteins allows least adsorption by proteins readily. These membranes can also encapsulate lipid molecules, such as cholesterol on their surfaces, being highly resistant to proteins. Other PEO-coated surfaces can also adsorb VLDL, LDL and HDL irreversibly over time and may cause platelets adhesion, resulting in thrombosis [93].

Many studies have suggested that proteins and enzymes are completely denatured after adsorption, while some sort of limited activity of probes, proteins and enzymes is retained in adsorbed state [90], particularly when these molecules are adsorbed from high concentration to reduce unfolding of 3D conformation induced by adsorption [89].

The study of retention time of enzyme activity through hydrophobic chromatography has revealed that the denaturation of any enzyme is dependent on different factors after adsorption: duration of adsorption over the surface of matrix, solvent, pH, temperature and the concentration of enzyme [91]. Cellular adhesion of platelets to the biomaterial surface can be associated with the specific amount of antifibrinogen interactions, signifying adsorbed fibrinogen in a totally different conformation on biomaterial surface. After adsorption of fibrinogen, a transitional state results, with reduced adhesion of antibodies and platelets, and plasma is displaced by changing amide II frequency. The binding of antibodies can be prevented by adding albumin the solvent [90, 94].

Another study of fibronectin on different biomaterial surfaces shows different results – fibronectin adsorbed to tissue culture-grade polystyrene-supported adsorption and cell adhesions. Otherwise, fibronectin adsorbed to regular polystyrene surface did not support cell adhesions, but albumin was added to aid the process [70]. The self-assembled monolayers (SAM) interact with other biomolecules to a greater extent, as the hydrophobic monolayers only support the process when these are rescued by albumins. Thus, albumin promotes cell adhesions and cell spreading by being adsorbed to different polymers and promotes the outgrowth of corneal epithelial cells. Albumin not only increases the adhesion of different biomaterial-coated surfaces but also decreases the competition between fibronectin molecules for the available sites [95]. Uniform biomaterial surfaces with different active functional groups show the usefulness of modulation of surface chemistry and biological properties by adsorbed molecules. The biomaterial surfaces containing terminal functional groups such as CH_3, COOH, OH and NH_2, adsorbed by fibronectin were found to cause varying cellular adhesions. The surface with high amount of CH_3 requires higher amount of fibronectin for cellular adhesion than OH surfaces [70, 89, 95].

2.6 Techniques to study adsorption via biomaterials

Various techniques have been developed to study mechanism of adsorption, molecular events, and to characterize properties of biomaterials. This section will elaborate the exclusive techniques to monitor the mechanism of adsorption and the molecules adsorbing on surface or already adsorbed molecules on the surface. Researchers focus on various aspects while studying the subject of biomaterials: kinetics of the reaction, thickness of the membrane, structure or conformation of adsorbed molecules, type of biomolecule, the operating forces between the biomolecule and matrix and the biocompatibility of both reactants. These aspects, somehow, urge them to innovate in the field of biomaterials. To study the conformation and structure of the adsorbed biomolecules, spectroscopy/spectrometry provides useful data. Infrared spectroscopy uses an IR beam, and a spectrum is developed by different IR absorbance peaks by different components. Different chemical bonds allow different vibrations and frequencies, and thus are absorbed by IR at different frequencies to allow them to generate an IR fingerprint, which, ultimately, provides information about conformation of adsorbed molecule [70]. IR spectrometry analyzes the bioactivity of biomolecules by detection of the IR spectra. The spectra differ according to the respective vibrational bands. Another advantage of this method is to cancel out the noise and collect the required vibrational bands only. The changes in concentration of any biomolecule can be measured over time, as the concentration of adsorbed molecule varies with time, on the surface. Another way to quantify the adsorbed molecule may be its loss in the bulk matrix, and the kinetics can be studied by plot of change in concentration of bulk versus time. The labeled fluorescent proteins, probes or fragments can be used to measure the radioactivity at the surface [70, 96].

Secondary ion mass spectrometry (SIMS) is significantly advantageous in studying adsorbed protein layers at a higher resolution, by bombardment of ions, causing emission of particles, which are analyzed for molecular data on its surface. Time-of-flight is coupled with SIMS to enhance the activity by free drifting and accelerating the secondary ions, allowing its increased separation and sensitivity. Modern mass spectrometry classifies the biomolecules like proteins into multiple components containing solutions, which can be applied to both in vivo and in vitro systems and are able to resolve a number of molecules of different molecular masses. X-ray photoelectron spectroscopy (XPS) is another technique to study the conformation of adsorbed proteins at the surface, irradiated by monochromatic X-rays, and released photoelectrons are captured to analyze [97] for molecules.

Attenuated total reflectance coupled with Fourier transform infrared spectroscopy, (ATR)-FTIR, provides information about the structure of adsorbed molecule by measurement of all the wavelengths at once; hence, making the biomaterial surface's cross-sectional area smaller for examination provides not only conformational analysis but also molecular and submolecular level data about molecules. Surface enhanced infrared adsorption spectroscopy (SEIRA), a modified form of IR spectroscopy, is an improved form used to study and recognize the adsorbed molecules like proteins, DNA and RNA. It has been used to study the interactions between biomolecules and biomaterial surfaces. SEIRA has improved the resolution and amplification of signal by two orders. It not only recognizes the biomolecules but also provides high levels of sensitivity, as improved analysis about changes in conformation of biomolecules predicts the change in functionality of adsorbed molecule [98].

The ELISA method, which measures the rate of protein adsorption in biological systems, is also another effective method. The targeted adsorbed proteins are detected and specified by specific antibodies in a recognizable conformation, by either providing information about its bioactivity at a certain site or by characterizing the presence of protein on targeted surface [90].

By utilizing polarized light, ellipsometry is used to examine the thickness of adsorbed molecule by a detector. The polarized light is received by a nondestructive surface at a definite angle and reflected back. Changes in reflected beam provide data about the interaction and thickness of the adsorbed proteins [99]. Surface plasmon resonance (SPR) is similar to ellipsometry and works by exciting the surface plasmons (rather than using polarized light). Neutron reflectivity is another technique to detect the adsorption rate by using a beam of neutrons directed at the surface and reflected back in a monitor to detect the reflected radiation profile. It predicts either the thickness of adsorbed biomolecule or its conformation It is risky to use, if the surface is radioactive and causes irradiation; due to its drawbacks, it is not widely used [15, 90].

Atomic force microscopy (AFM) has allowed the recognition of biomolecules in their native form and predicted their three-dimensional structure. A cantilevered tip works to function on the surface with a very thin microscale. Recently, AFM coupled with scanning tunneling microscope (STM) has described one adsorbed particle protein accurately, by measuring quantum current between the surface and cantilevered tip. Subsequently, the excavating electrons move among the cantilevered tip or conducting surface, and an automated image is generated by capturing every single change in height of scanned area. STM-AFM is now being widely used to distinguish the conformational changes in metallo-proteins [100].

Mass spectrometry using matrix-assisted laser desorption ionization time-of-flight (MALDI-ToF) provides limitless mass resolution at even low detection limits. A pulse light source is used to irradiate the adsorbed layer, the measured ions from the release; mass-to-charge ratio is created, compared with available database and analyzed to determine particular structure by comparison with available data [101]. This method can observe a mass range and resolve larger structures. The threshold point for sensitivity of any biomolecule relies heavily on the individual spectrometer used. The researchers are focusing on an innovation in MALDI by digesting the adsorbed molecules with certain enzymes and blotting the solution on matrix plate.

The analysis by a mass spectrometer with a lower range would result in detection of larger fragments and enhanced selectivity [102].

MALDI-ToF/MS 2D gel electrophoresis can be used to isolate adsorbed proteins from the surface in the form of specific bands. MALDI-ToF/MS has potentially limited use, as proteins cannot be recognized at low concentration, and protein can be denatured during MALDI. Some advances in mass spectrometry, MALDI and IR will resolve these problems by using tandem approaches and improved protein processors [89]. Transmission electron microscope (TEM), scanning electron microscopy (SEM) and solid-state crystallography can be used to study the conformation of adsorbed molecules too. There are, however, some drawbacks: they require high concentration of adsorbed molecules, altering both the surface and chemistry, and distort the layer structure on it. These methods require the molecule in dehydrated form for analysis; the dehydrated form can alter the chemistry of both surface and molecule [89, 101].

Tandem scanning probe microscopy technique suggests the details of every single biomolecule adsorbed on the surface of biomaterial [103]. The advancements in spectroscopy fields may offer a new gateway to study the conformation of every single atom at cellular levels and would be a great breakthrough. Recent advancements have allowed the improved version of sensitivities and shown the tremendous potential of adsorption by biomaterial surfaces. To progress to higher resolutions of spectrometry, scientists are studying a few surface treatments by specifically using certain components in the matrix to protect the sample from denaturation and aggregation during the reaction, and it has improved the vision of the story.

2.7 Conclusion

As shown in this chapter, the process of adsorption to biological material surfaces under various conditions provides fundamental information and understanding of cellular adhesions, biological responses following biocompatibility and molecular events. In various studies, scientists have reported variations in conformation of biomolecules and changes in bioactivity after adsorption; however, molecular events relating to the biomaterial surface and its chemistry, conformational changes and bioactivity are still not fully elaborated. Nanomaterials is an untapped field that will help study the biomaterial phenomenon at nanoscale. Using different techniques and coupling them with nanomaterials chemistry will add valued data and an improved understanding about biomolecule-surface interactions. By improving understanding of adsorption, Nanochemistry will allow an enhanced control over specific adsorption (by successive cellular adhesions and its functions relevant to tissue implantation), and it will also lead to design and formulation of improved biomaterials for clinical applications.

References

[1] Young, B., W. Pitt, and S. Cooper, *Protein adsorption on polymeric biomaterials I. Adsorption isotherms*. Journal of Colloid and Interface Science, 1988. **124**(1): p. 28–43.

[2] Park, J. and R.S. Lakes, *Biomaterials: an introduction*. 2007: Springer Science & Business Media.

[3] Naidu, N.A., Wadher, K. and Umekar, M., 2021. An overview on biomaterials: Pharmaceutical and biomedical applications. *Journal of Drug Delivery and Therapeutics*, **11**(1-s), pp.154–161.

[4] Chen, F.M. and Liu, X., 2016. Advancing biomaterials of human origin for tissue engineering. *Progress in polymer science*, **53**, pp.86–168.

[5] Lam, M.T. and Wu, J.C., 2012. Biomaterial applications in cardiovascular tissue repair and regeneration. *Expert review of cardiovascular therapy*, **10**(8), pp.1039–1049.

[6] Schmidt, D.R., H. Waldeck, and W.J. Kao, *Protein adsorption to biomaterials*, in *Biological interactions on materials surfaces*. 2009, Springer. p. 1–18.

[7] Young, B., W. Pitt, and S. Cooper, *Protein adsorption on polymeric biomaterials: II. Adsorption kinetics*. Journal of Colloid and Interface Science, 1988. **125**(1): p. 246–260.

[8] Park, T.H. and Shuler, M.L., 2003. Integration of cell culture and microfabrication technology. *Biotechnology progress*, **19**(2), pp.243–253.

[9] Ratner, B.D. and G. Zhang, *A history of biomaterials*, in *Biomaterials science*. 2020, Elsevier. p. 21–34.

[10] Ratner, B.D., et al., *Biomaterials science: an introduction to materials in medicine*. 2004: Elsevier.

[11] Kamatar, A., Gunay, G. and Acar, H., 2020. Natural and synthetic biomaterials for engineering multicellular tumor spheroids. *Polymers*, **12**(11), p.2506.

[12] Beena, V., et al., *Photocatalytic dye degradation and biological activities of Cu-doped ZnSe nanoparticles and their insights*. Water, 2021. **13**(18): p. 2561.

[13] Bibi, S., et al., *Synthesis of Cr doped LiMnPO4 cathode materials and investigation of their dielectric properties*. International Journal of Energy Research, 2021.

[14] Fan, H., et al., *Adsorption properties and mechanisms of novel biomaterials from banyan aerial roots via simple modification for ciprofloxacin removal*. Science of The Total Environment, 2020. **708**: p. 134630.

[15] Kim, J., *Systematic approach to characterize the dynamics of protein adsorption on the surface of biomaterials using proteomics*. Colloids and Surfaces B: Biointerfaces, 2020. **188**: p. 110756.

[16] Othman, Z., Pastor, B.C., van Rijt, S. and Habibovic, P., 2018. Understanding interactions between biomaterials and biological systems using proteomics. *Biomaterials*, **167**, pp.191–204.

[17] Zhou, T., Zhu, Y., Li, X., Liu, X., Yeung, K.W., Wu, S., Wang, X., Cui, Z., Yang, X. and Chu, P.K., 2016. Surface functionalization of biomaterials by radical polymerization. *Progress in Materials Science*, **83**, pp.191–235.

[18] Holmes, C. and Tabrizian, M., 2015. Surface functionalization of biomaterials. *In Stem cell biology and tissue engineering in dental sciences* (pp. 187–206). Academic Press.

[19] Kaneko, H., Kamiie, J., Kawakami, H., Anada, T., Honda, Y., Shiraishi, N., Kamakura, S., Terasaki, T., Shimauchi, H. and Suzuki, O., 2011. Proteome analysis of rat serum proteins adsorbed onto synthetic octacalcium phosphate crystals. *Analytical biochemistry*, **418**(2), pp.276–285.

[20] Araújo-Gomes, N., Romero-Gavilán, F., Sánchez-Pérez, A.M., Gurruchaga, M., Azkargorta, M., Elortza, F., Martinez-Ibañez, M., Iloro, I., Suay, J. and Goni, I., 2018. Characterization of serum proteins attached to distinct sol–gel hybrid surfaces. *Journal of Biomedical Materials Research Part B: Applied Biomaterials*, **106**(4), pp.1477–1485.

[21] Pitt, W.G., K. Park, and S.L. Cooper, *Sequential protein adsorption and thrombus deposition on polymeric biomaterials*. Journal of colloid and interface science, 1986. **111**(2): p. 343–362.

[22] Bouchard, B.A. and Tracy, P.B., 2003. The participation of leukocytes in coagulant reactions. *Journal of Thrombosis and Haemostasis*, **1**(3), pp.464–469.

[23] Lomer, M.C., Thompson, R.P. and Powell, J.J., 2002. Fine and ultrafine particles of the diet: influence on the mucosal immune response and association with Crohn's disease. *Proceedings of the Nutrition Society*, **61**(1), pp.123–130.

[24] Latour, R.A., 2005. Biomaterials: protein-surface interactions. *Encyclopedia of biomaterials and biomedical engineering*, **1**, pp.270–284.

[25] Wronska, M.A., O'Connor, I.B., Tilbury, M.A., Srivastava, A. and Wall, J.G., 2016. Adding functions to biomaterial surfaces through protein incorporation. *Advanced Materials*, **28**(27), pp.5485–5508.

[26] Pataranutaporn, P., T. Ingalls and E. Finn. Biological HCI: towards integrative interfaces between people, computer, and biological materials. In Extended Abstracts of the 2018 CHI Conference on Human Factors in Computing Systems (1–6). 2018 April.

[27] Dunlop, J.W., R. Weinkamer and P. Fratzl. Artful interfaces within biological materials. Materials Today: Proceedings, 2011. **14**(3): p. 70–78.

[28] Lee, H.P. and A.K. Gaharwar. Light-responsive inorganic biomaterials for biomedical applications. Advanced Science, 2020. **7**(17): p. 2000863.

[29] Vincent, J., *Structural biomaterials*. 2012: Princeton University Press.

[30] Zhang, X.Z., et al., *Synthesis, characterization and wettability of Cu-Sn alloy on the Si-implanted 6H-SiC*. Coatings, 2020. **10**(9): p. 906.

[31] Patwardhan, S.V., G. Patwardhan, and C.C. Perry, *Interactions of biomolecules with inorganic materials: principles, applications and future prospects*. Journal of Materials Chemistry, 2007. **17**(28): p. 2875–2884.

[32] Aiyelabegan, H.T. and E. Sadroddiny, *Fundamentals of protein and cell interactions in biomaterials*. Biomedicine & Pharmacotherapy, 2017. **88**: p. 956–970.

[33] Chinn, J.A. and S. Slack, *Biomaterials: Protein – Surface Interactions*. The Biomedical Engineering Handbook:, 2013.

[34] Hasirci, V. and N. Hasirci, *Fundamentals of biomaterials*. 2018: Springer.

[35] Perry, C.C., S.V. Patwardhan, and O. Deschaume, *From biominerals to biomaterials: The role of biomolecule–mineral interactions*. Biochemical Society Transactions, 2009. **37**(4): p. 687–691.

[36] Horbett, T.A., *Fibrinogen adsorption to biomaterials*. Journal of Biomedical Materials Research Part A, 2018. **106**(10): p. 2777–2788.

[37] Latour, R.A., *Biomaterials: protein-surface interactions*. Encyclopedia of Biomaterials and Biomedical Engineering, 2005. **1**: p. 270–278.

[38] Hlady, V. and J. Buijs, *Protein adsorption on solid surfaces*. Current Opinion in Biotechnology, 1996. **7**(1): p. 72–77.

[39] Ngo, B.K.D. and M.A. Grunlan, *Protein resistant polymeric biomaterials*. 2017, ACS Publications.

[40] Bruck, S.D., *Properties of biomaterials in the physiological environment*. 2019: CRC Press.

[41] García, A.J., *Surface modification of biomaterials*. Foundations of Regenerative Medicine: Clinical and Therapeutic Applications, 2009: p. 368–378.

[42] Cass, A., T. Cass, and F.S. Ligler, *Immobilized biomolecules in analysis: a practical approach*. Vol. 198. 1998: Practical Approach (Paperback).

[43] Elnashar, M.M., *Immobilized molecules using biomaterials and nanobiotechnology*. Journal of Biomaterials and Nanobiotechnology, 2010. **1**(01): p. 61.

[44] Rao, S.S. and J. Winter, *Adhesion molecule-modified biomaterials for neural tissue engineering*. Frontiers in Neuroengineering, 2009. **2**: p. 6.
[45] Hoffman, A., *Immobilization of biomolecules and cells on and within polymeric biomaterials*, in *Biologically Modified Polymeric Biomaterial Surfaces*. 1992, Springer. p. 61–65.
[46] Ahuja, T., I.A. Mir, and D. Kumar, *Biomolecular immobilization on conducting polymers for biosensing applications*. Biomaterials, 2007. **28**(5): p. 791–805.
[47] Agarwal, R. and A.J. García, *Surface modification of biomaterials*, in *Principles of Regenerative Medicine*. 2019, Elsevier. p. 651–660.
[48] Ji, W., et al., *Local delivery of small and large biomolecules in craniomaxillofacial bone*. Advanced Drug Delivery Reviews, 2012. **64**(12): p. 1152–1164.
[49] Duan, B., et al., *Surface modification of three-dimensional Ca-P/PHBV nanocomposite scaffolds by physical entrapment of gelatin and its in vitro biological evaluation*. Frontiers of Materials Science, 2011. **5**(1): p. 57–68.
[50] Costa, D., P.A. Garrain, and M. Baaden, *Understanding small biomolecule-biomaterial interactions: A review of fundamental theoretical and experimental approaches for biomolecule interactions with inorganic surfaces*. Journal of Biomedical Materials Research Part A, 2013. **101**(4): p. 1210–1222.
[51] Wang, Q., et al., *Computer simulation of biomolecule–biomaterial interactions at surfaces and interfaces*. Biomedical Materials, 2015. **10**(3): p. 032001.
[52] Hu, B., et al., *Recent advances in chitosan-based layer-by-layer biomaterials and their biomedical applications*. Carbohydrate Polymers, 2021: p. 118427.
[53] Parikh, A.N. and D.L. Allara, *Self-assembled monolayers: a versatile tool for biofunctionalization of surfaces*. Handbook of Biofunctional Surfaces. Singapore: Pan Stanford Publishing Pte. Ltd, 2013: p. 1–29.
[54] Prashar, D., *Self assembled monolayers-a review*. Int. J. ChemTech Res, 2012. **4**(1): p. 258–265.
[55] Keeney, M., et al., *Nanocoating for biomolecule delivery using layer-by-layer self-assembly*. Journal of Materials Chemistry B, 2015. **3**(45): p. 8757–8770.
[56] Alkekhia, D., P.T. Hammond, and A. Shukla, *Layer-by-layer biomaterials for drug delivery*. Annual Review of Biomedical Engineering, 2020. **22**: p. 1–24.
[57] Pei, R., et al., *Assembly of alternating polycation and DNA multilayer films by electrostatic layer-by-layer adsorption*. Biomacromolecules, 2001. **2**(2): p. 463–468.
[58] Elaissari, A., *Colloidal biomolecules, biomaterials, and biomedical applications*. Vol. 116. 2003: CRC Press.
[59] Love, R.J. and K.S. Jones, *The recognition of biomaterials: pattern recognition of medical polymers and their adsorbed biomolecules*. Journal of Biomedical Materials Research Part A, 2013. **101**(9): p. 2740–2752.
[60] Custódio, C.A., R.L. Reis, and J.F. Mano, *Engineering biomolecular microenvironments for cell instructive biomaterials*. Advanced Healthcare Materials, 2014. **3**(6): p. 797–810.
[61] von der Mark, K., et al., *Nanoscale engineering of biomimetic surfaces: cues from the extracellular matrix*. Cell and Tissue Research, 2010. **339**(1): p. 131–153.
[62] Shin, H., S. Jo, and A.G. Mikos, *Biomimetic materials for tissue engineering*. Biomaterials, 2003. **24**(24): p. 4353–4364.
[63] Rahmany, M.B. and M. Van Dyke, *Biomimetic approaches to modulate cellular adhesion in biomaterials: A review*. Acta Biomaterialia, 2013. **9**(3): p. 5431–5437.
[64] Vogler, E.A., *Structure and reactivity of water at biomaterial surfaces*. Advances in Colloid and Interface Science, 1998. **74**(1–3): p. 69–117.

[65] Kalasin, S. and M.M. Santore, *Non-specific adhesion on biomaterial surfaces driven by small amounts of protein adsorption*. Colloids and Surfaces B: Biointerfaces, 2009. **73**(2): p. 229–236.
[66] Rabinow, B., et al., *Biomaterials with permanent hydrophilic surfaces and low protein adsorption properties*. Journal of Biomaterials Science, Polymer Edition, 1995. **6**(1): p. 91–109.
[67] Ratner, B., A. Johnston, and T. Lenk, *Biomaterial surfaces*. Journal of biomedical materials research, 1987. **21**(A1 Suppl): p. 59–89.
[68] Fabrizius-Homan, D. and S. Cooper, *A comparison of the adsorption of three adhesive proteins to biomaterial surfaces*. Journal of Biomaterials Science, Polymer Edition, 1992. **3**(1): p. 27–47.
[69] Kasemo, B. and J. Lausmaa, *Surface science aspects on inorganic biomaterials*. CRC Critical Reviews of Clinical Neurobiology (United States), 1986. **4**.
[70] Hasan, A., S.K. Pattanayek, and L.M. Pandey, *Effect of functional groups of self-assembled monolayers on protein adsorption and initial cell adhesion*. ACS Biomaterials Science & Engineering, 2018. **4**(9): p. 3224–3233.
[71] Arima, Y. and H. Iwata, *Effect of wettability and surface functional groups on protein adsorption and cell adhesion using well-defined mixed self-assembled monolayers*. Biomaterials, 2007. **28**(20): p. 3074–3082.
[72] Arima, Y. and H. Iwata, *Effects of surface functional groups on protein adsorption and subsequent cell adhesion using self-assembled monolayers*. Journal of Materials Chemistry, 2007. **17**(38): p. 4079–4087.
[73] Martins, M.C.L., B.D. Ratner, and M.A. Barbosa, *Protein adsorption on mixtures of hydroxyl- and methyl-terminated alkanethiols self-assembled monolayers*. Journal of Biomedical Materials Research Part A: An Official Journal of The Society for Biomaterials, The Japanese Society for Biomaterials, and The Australian Society for Biomaterials and the Korean Society for Biomaterials, 2003. **67**(1): p. 158–171.
[74] Latour Jr, R.A., *Molecular modeling of biomaterial surfaces*. Current Opinion in Solid State and Materials Science, 1999. **4**(4): p. 413–417.
[75] Hoven, V.P., et al., *Surface-charged chitosan: Preparation and protein adsorption*. Carbohydrate Polymers, 2007. **68**(1): p. 44–53.
[76] Kingshott, P. and H.J. Griesser, *Surfaces that resist bioadhesion*. Current Opinion in Solid State and Materials Science, 1999. **4**(4): p. 403–412.
[77] Amiji, M. and K. Park, *Prevention of protein adsorption and platelet adhesion on surfaces by PEO/PPO/PEO triblock copolymers*. Biomaterials, 1992. **13**(10): p. 682–692.
[78] Bergström, K., et al., *Reduction of fibrinogen adsorption on PEG-coated polystyrene surfaces*. Journal of Biomedical Materials Research, 1992. **26**(6): p. 779–790.
[79] Barbosa, J.N., et al., *Adhesion of human leukocytes on mixtures of hydroxyl-and methyl-terminated self-assembled monolayers: Effect of blood protein adsorption*. Journal of Biomedical Materials Research Part A: An Official Journal of The Society for Biomaterials, The Japanese Society for Biomaterials, and The Australian Society for Biomaterials and the Korean Society for Biomaterials, 2010. **93**(1): p. 12–19.
[80] Aghajani, M. and F. Esmaeili, *Anti-biofouling Assembly Strategies for Protein & Cell Repellent Surfaces: A mini-review*. Journal of Biomaterials Science, Polymer Edition, 2021(just-accepted): p. 1–21.
[81] Gombotz, W.R., et al., *Protein adsorption to poly (ethylene oxide) surfaces*. Journal of Biomedical Materials Research, 1991. **25**(12): p. 1547–1562.
[82] Zhang, M., T. Desai, and M. Ferrari, *Proteins and cells on PEG immobilized silicon surfaces*. Biomaterials, 1998. **19**(10): p. 953–960.

[83] Hoffman, A.S., *The early days of PEG and PEGylation (1970s–1990s)*. Acta Biomaterialia, 2016. **40**: p. 1–5.

[84] Kwon, H.J., et al., *Zwitterionic sulfobetaine polymer-immobilized surface by simple tyrosinase-mediated grafting for enhanced antifouling property*. Acta Biomaterialia, 2017. **61**: p. 169–179.

[85] Batool, F., et al., *Modulation of immune-inflammatory responses through surface modifications of biomaterials to promote bone healing and regeneration*. Journal of Tissue Engineering, 2021. **12**: p. 20417314211041428.

[86] Mohan, T., et al., *Highly Protein Repellent and Antiadhesive Polysaccharide Biomaterial Coating for Urinary Catheter Applications*. ACS Biomaterials Science & Engineering, 2019. **5**(11): p. 5825–5832.

[87] Esmaeilzadeh, P., et al., *Stimuli-responsive multilayers based on thiolated polysaccharides that affect fibroblast cell adhesion*. ACS Applied Materials & Interfaces, 2018. **10**(10): p. 8507–8518.

[88] Asker, D., et al., *Non-eluting, surface-bound enzymes disrupt surface attachment of bacteria by continuous biofilm polysaccharide degradation*. Biomaterials, 2018. **167**: p. 168–176.

[89] Othman, Z., et al., *Understanding interactions between biomaterials and biological systems using proteomics*. Biomaterials, 2018. **167**: p. 191–204.

[90] Visalakshan, R.M., et al., *Biomaterial surface hydrophobicity-mediated serum protein adsorption and immune responses*. ACS Applied Materials & Interfaces, 2019. **11**(31): p. 27615–27623.

[91] Mariani, E., et al., *Biomaterials: foreign bodies or tuners for the immune response?* International Journal of Molecular Sciences, 2019. **20**(3): p. 636.

[92] Romero-Gavilán, F., et al., *The effect of strontium incorporation into sol-gel biomaterials on their protein adsorption and cell interactions*. Colloids and Surfaces B: Biointerfaces, 2019. **174**: p. 9–16.

[93] Brash, J.L., et al., *The blood compatibility challenge. Part 2: Protein adsorption phenomena governing blood reactivity*. Acta Biomaterialia, 2019. **94**: p. 11–24.

[94] Xu, L.C. and C.A. Siedlecki, *Protein adsorption, platelet adhesion, and bacterial adhesion to polyethylene-glycol-textured polyurethane biomaterial surfaces*. Journal of Biomedical Materials Research Part B: Applied Biomaterials, 2017. **105**(3): p. 668–678.

[95] Xu, L.-n., et al., *Low density lipoprotein adsorption on a titanium surface and its effect on osteoblast behaviors*. RSC Advances, 2019. **9**(32): p. 18589–18598.

[96] Mei, S., et al., *Influences of tantalum pentoxide and surface coarsening on surface roughness, hydrophilicity, surface energy, protein adsorption and cell responses to PEEK based biocomposite*. Colloids and Surfaces B: Biointerfaces, 2019. **174**: p. 207–215.

[97] Höhn, S., S. Virtanen, and A.R. Boccaccini, *Protein adsorption on magnesium and its alloys: A review*. Applied Surface Science, 2019. **464**: p. 212–219.

[98] Tanaka, M., S. Morita, and T. Hayashi, *Role of interfacial water in determining the interactions of proteins and cells with hydrated materials*. Colloids and Surfaces B: Biointerfaces, 2020: p. 111449.

[99] Guzman, G., et al., *Transport-limited adsorption of plasma proteins on bimodal amphiphilic polymer co-networks: real-time studies by spectroscopic ellipsometry*. Langmuir, 2017. **33** (11): p. 2900–2910.

[100] Hasan, A. and L.M. Pandey, *Implications of the Nanoscopic Surface Modification on the Protein Adsorption and Cell Adhesion*. Biobased Nanotechnology for Green Applications, 2021: p. 423–460.

[101] Wang, L., et al., *Zwitterionic Janus Dendrimer with distinct functional disparity for enhanced protein delivery*. Biomaterials, 2019. **215**: p. 119233.

[102] Tobieson, L., et al., *Proteomic investigation of protein adsorption to cerebral microdialysis membranes in surgically treated intracerebral hemorrhage patients-a pilot study*. Proteome Science, 2020. **18**(1): p. 1–12.

[103] Buck, E., et al., *Protein Adsorption on Surfaces Functionalized with COOH Groups Promotes Anti-inflammatory Macrophage Responses*. ACS Applied Materials & Interfaces, 2021. **13**(6): p. 7021–7036.

Sadaf Tariq, Awais Ahmad, Rafael Luque, Mabkhoot Alsaiari, Farid A. Harraz

3 Recent trends in biomaterials for photodegradation of dyes

Abstract: Synthetic dyes are highly carcinogenic and can cause life-threatening problems. These dyes are part of products of daily use, including textile mills, leather industries, common accessories and furniture. Dyes are released in the environment and water as a major effluent, due to different anthropogenic and industrial activities. The accumulation of nondegradable dyes in environment poses serious health threats. Different conventional methods have been reported so far as being used to degrade the dyes from environment. Photodegradation is one of the most promising, cost-effective and facile methods to remove dyes. Recently, biomaterials have been used as a new frontier to degrade dyes by heterogeneous photodegradation, and it has appeared as the most emerging technology due to its cost effectiveness. This chapter deals with photodegradation of dyes via biomaterials. The photodegradation properties with recent advances and trends are discussed. It includes a brief understanding of photodegradation, the potential use of biomaterials and their composites in the photodegradation of synthetic and toxic dyes. Furthermore, we conclude with further prospects to improve the efficiency of biomaterials used for removal of dyes.

Keywords: Dyes, biomaterials, photodegradation, toxic components, water, environment

3.1 Introduction

Synthetic dyes are annually synthesized and used in textile, leather, nutraceutical, pharmaceutical, cosmetic and paper industries. More than 100,000 synthetic dyes are commercially available, with an annually increasing synthesis rate, most of which contaminates the environment and water. Even trace amounts of dyes can contaminate the environment in an undesirable manner. These contaminants can cause serious health threats to aquatics and also distort human health [1–3]. Colored wastewater discharge and inefficient treatment methodologies can cause serious environmental concerns. So far, different conventional strategies have been developed to remove dyes from the environment, including adsorption, biodegradation, coagulation, photodegradation and adsorption. Among these approaches, photodegradation can be used to degrade and remove dyes through different biomaterials and their derivatives [4–6].

https://doi.org/10.1515/9783110768749-003

In this chapter, the photodegradation of dyes based on different biomaterials and their derivatives will be discussed in subsequent sections. Different biomaterials-assisted photocatalysts can be prepared using biotemplates for their enhanced photodegradation of dyes. The water solubility, coordination of biomaterials with dyes and cooperativity of components for biodegradation lead to efficient binding of dyes and biomaterials [7–9]. The as-obtained biohybrids possess enhanced activity by promoting light absorption and the ability to accept electrons generated due to the absorption of photons of light, thereby yielding superoxide radicals [10, 11]. The scheme of reaction mechanism can be illustrated as shown below:

$$\mathrm{Dye} + h\nu \rightarrow \mathrm{Dye}^*$$

$$\mathrm{Dye}^* + \mathrm{TiO_2} \rightarrow \mathrm{Dye}^+$$

$$\mathrm{Biomaterial(e)} + \mathrm{O_2} \rightarrow \mathrm{biomaterial(e)} + \mathrm{O}^{2-}$$

$$\bullet\,\mathrm{O}^{2-} + \mathrm{biomaterial(e)} + 2\mathrm{H}^+ \rightarrow \mathrm{H_2O_2} + \mathrm{biomaterial}$$

$$\bullet\,\mathrm{H_2O_2} + \mathrm{biomaterial(e)} \rightarrow \mathrm{OH{+}OH^-} + \mathrm{biomaterial}$$

Biomaterial-assisted synthesis of biohybrids can be used as an attractive and promising method to degrade dyes. The carbon-rich biomaterials have been derived from biomass to synthesize biochars under pyrolysis, with limited oxygen supply, to undergo carbonization [12–14]. Usually, cellulose, lignin and hemicellulose are used as biomass to assist in the preparation of biomaterials. The modified and functionalized biomaterials find their applications in environmental remediation. Some bio-carbonaceous materials, including biotemplates, biochars, biomolecules and bioinspired molecules are replacing conventional materials, due to their versatile nature in removing different pollutants such as antibiotics, aromatic dyes and organic pollutants [15–17].

The subsequent sections of this chapter briefly focus on the latest and most recent breakthroughs in use of biomaterials for photodegradation of dyes, by using different green chemistry principles for safer and bioinspired remediation of pollutants. Certain plant materials, biomolecules and biohybrids may open an avenue for the development of biomaterials for environmental remediation. Furthermore, we conclude with future prospects for nanotechnology and biomaterials for enhanced applications.

3.2 Dye molecules

Owing to the solubility, stability and activity of photosensitive dyes during homogeneous photocatalysis, ruthenium analogues and its complexes are commonly used.

Recently, the cationic complex of Ru(II) and tris(2,2′-bipyrazyl) has been used to initiate the redox reaction, by acting as an artificial cofactor to continuously reduce and lower the catalysis of their equivalents [18, 19]. Furthermore, the cationic complex of tris(2,2′-bipyridyl) and Ru(II) have been applied to remove the electron–hole pairs from water molecules, under visible light. The Ru-based photocatalysts enhance the optical properties by making available three phases of Ru, by linking them with hydrophobic interactions [20]. The hybrids and complexes of Ru facilitate the delocalization of excitons by empowering the interaction between ligands and nanostructures. Additionally, the efficiency of these complexes cannot only be correlated to the intrinsic absorption of light by Ru as photocatalyst, but also affect the transfer of energy from complexes to the main chemical reaction system, which proves the consequence of the synergetic effect between photons and complexes [6, 21].

Owing to the high efficacy of quantum effect of dyes, the xanthene dyes capture visible light. The eosin Y is a dye molecule that is known for its implication as a photocatalyst in hybrid system. The photocatalyst finds its application in regeneration of NADH at an exceptionally high rate. The photocatalytic property can be attributed to the reversible interactions between eosin Y and metal center. The Rh complexes are important in this regard too, by acting as a potential gradient to transfer electrons in a specific vicinity [21, 22]. Therefore, eosin Y is capable of photogeneration of NADPH, which, thereby facilitates the cytochrome P450 to catalyze the dealkylation of 7-ethoxycoumarin. The light-harvesting capacity can be enhanced by core shell-nanostructures, as plasmonic nanohybrids of eosin Y have been reported. The fine encapsulation enhances the local surface plasmon resonance to amplify the absorption of photons and generates the charge carriers. Thioflavin T and benzothiazole dye possess specific binding affinity for β-sheets amyloids. Owing to these properties, light-harvesting nanofibrils have been reported by using insulin as precursor, to improve the photocatalytic activity and extend the excited half-life of electrons of ThT [6, 7, 20].

3.3 Biomaterials for dye degradation

Owing to the large-scale use of dyes in many industries, certain technologies and materials have been designed to remove/degrade the dye molecules from aqueous solution, so far. Thousands of articles have reported the use of different techniques and materials to remove dyes and many recommended cost-effective and eco-friendly materials for degradation of dyes, indicating their interest to search for new techniques. The most recent techniques in dye degradation techniques are adsorption, advanced oxidation processes (AOPs), coagulation/flocculation, electrochemical/biological treatment, membrane filtration, and many others. Albeit efforts made so far for development of an effective technology for degradation of dyes, many of them pose a few challenges for their general use [23, 24]. Most of the bottlenecks of these technologies are their high cost, labor involvement, environment hazardous nature and complexity. Most of them use precursors which are nonbiodegradable and cause pollution, themselves. Moreover, chemical-based methods are also avoided due to their multistep synthesis and nature of chemicals. Therefore, researchers are keen to develop bioinspired or biomaterial-based natural polymers, such as cellulose, chitin, chitosan and other biomolecules. Recently, different bioinspired materials have been developed and employed for the photodegradation of dyes [25–27].

The use of biomaterials in this regard is gaining importance due to their abundance, ease of availability, low cost, tailored surface properties (surface area and porosity), ease of handling and environment-friendly nature [11]. Biomaterials possess advantages over synthetic materials and chemicals, as they are biodegradable and do not pose any side effects on living organisms. Biomaterials are highly biocompatible and are derived from renewable sources, making them one of the most sustainable and suitable alternatives, over other materials. Since some functional classes are present, such as hydroxyl, amine, carboxyl and amide, it becomes easy to functionalize their surfaces, without affecting the biological and chemical properties of the biomaterial. Furthermore, the modified biomaterials exhibit increased biocompatibility, robustness, flexibility, crystallinity, surface area, size modifications, shape modifications, thermal stability, chemical stability and rigidity [11, 28].

Biomaterials can also be fabricated with other materials and structures, such as spherical, nanofilms, hydrogen bead, nanofibers, scaffolds and nanoparticles. However, recent advances in biomaterials are also accompanied by a variety of bottlenecks, such as contamination, fluctuations in the chemical and structural properties and addition of other impurities during the purification process [29]. This chapter summarizes all the current developments and recent advances of biomaterials and bionanomaterials with their major advantages and limitations.

3.3.1 Photocatalytic degradation of dyes by biopolymers

Biopolymers can be used to degrade dyes, due to their quick and facile recovery from the reaction system. Certain biopolymers, including cellulose and chitosan, act as support for metal oxide-containing photocatalysts, by providing a high surface area for strong adsorption. Owing to the high surface area, biopolymers reduce the reaction intermediates during photodegradation of dyes [27]. Additionally, chitosan and cellulose afford their reusability, as they can be recycled during photocatalysis. Photodegradation by using biopolymers is different in the manner that the photons are involved to induce photocatalysis during reaction. Due to photons, the oxygen molecules accept electrons generated by the photodegradation of dyes

[28]. Biopolymers have exhibited rapid and facile degradation of dyes under radiations of visible light. The concept of eco-friendly and cost-effective biopolymers for photodegradation of dyes was initially developed, a decade ago. Recently, green chemistry approach is being employed using nanomaterials for photocatalysis of dyes and other organic pollutants. Nanotechnology is now popular and explored for bioremediation, for its high performances and next-generation approaches. The reactive textile dyes can be removed by breaking the electron affinity between the lone pair of electrons and chelating agents [30, 31].

The nanoparticles interact with the backbone of the polymeric chain and distort the ionic nature to destabilize the interactions between them. Also, biopolymeric membrane-integrated metal oxides nanomaterials have attracted the attention of researchers for their better activity, selectivity, reusability and hydrophilicity, even at high temperatures [26, 32]. Gallic acid as biomaterial can be integrated into metal oxide nanomaterials and is now used in photodegradation of reactive dyes. The nanomaterial TiO_2/chitosan–acrylic acid integration has been explored for its photodegradation of Malachite green dye. The metal oxide and biomaterial integration such as Fe_3O_4/chitosan has been used for degradation of X-3B dye, under visible light [33]. The carbon black-cellulose acetate/ZnO is also under research for its degradation properties, and it has been reported to degrade Congo red, methyl orange and methylene blue dyes [34]. MnO_2/cellulose is another biomaterial-based metal oxide used to degrade indigo carmine [35]. The ZnO/cellulose and ZnO/cellulose nanofibers based on green chemistry approach have been successfully used to degrade methylene blue and Rhodamine B dyes [36]. Figure 3.1 illustrates the properties and analysis techniques used to describe biopolymeric nanomaterials.

Different characterization analysis can be used to explore different physical, structural, configurational, compositional and chemical properties of as-synthesized biomaterials, other composites and hybrids. These properties specifically determine the surface chemistry, functionalization, arrangement, dispersibility, agglomerations, functional groups, elemental composition, stability and activity and so on [17, 37]. The most commonly used techniques are microscopic [scanning electron microscopy (SEM), transmission electron microscopy (TEM), atomic absorption spectroscopy (AAS), atomic force microscopy (AFM) and dynamic light scattering (DLS)] and spectroscopic techniques [Fourier transform infrared spectroscopy (FTIR), X-ray photoelectron spectroscopy (XPS), UV–visible, zeta-potential, energy-dispersive X-ray spectroscopy (EDAX), and X-ray diffractometry (XRD)]. AAS is a very powerful characterization tool used to determine the concentrations of certain elements in biomaterials. XRD predicts the crystallinity and amorphous nature of any biomaterial. EDAX is the most commonly used method for elemental analysis in bionanomaterials. XPS provides information about the oxidation state of atom. FTIR provides information about the presence of functional groups in the bionanomaterials. SEM is used for determination of properties of the structure, surface area,

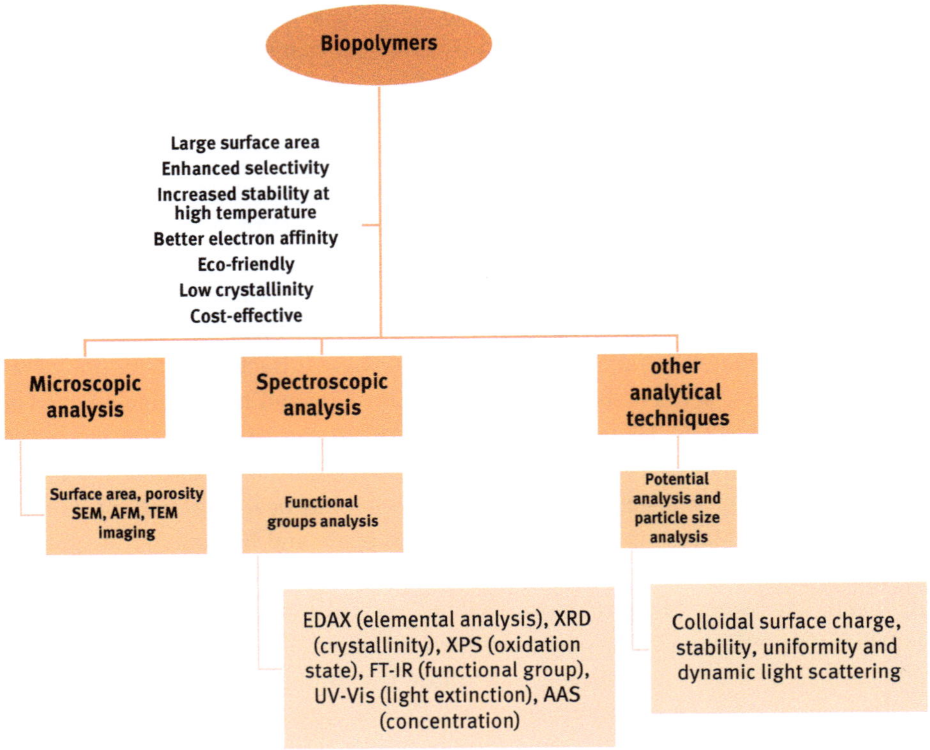

Figure 3.1: Properties and analysis of biopolymers.

and particle diameter. TEM is also an advanced technique that provides information about surface charge and particle size along with AFM [30, 38].

3.3.1.1 Synthesis of biopolymeric nanomaterials

Dyes absorb or emit light to produce color in visible light due to their chromogenic/chromophoric nature. Usually, dyes contain auxochromes (bridging) and solubilizing groups (reactive) anchor to the substrates with different functional groups through physical adsorption or covalent fixation. Typically, the aromatic structure of dyes contains benzene, naphthalene and anthracene rings. The aromatic rings carry auxochromic binding groups with delocalized electrons to form a conjugation. Certain functional groups affect the binding capacity of dyes, including azo, methine, ethylene, carbon-sulfur, carbonyl, nitro, nitroso, carbon-nitrogen, crinoid, carboxyl, amino, sulfonate and hydroxyl groups. In the production of cellulosic fibers, the reactive functional groups such as fluoro, chloro, methyl sulfone and nicotinyl of dyes become covalently bound to the substrate. The leaving groups of dyes and the substrates react through nucleophilic substitution reaction and get activated by adjacent nitrogen atoms of triazine ring [29, 39–41].

Conventionally, the polymeric nanomaterials can be synthesized by chemical methods; which are mostly toxic for their instability in environment, whereas the biopolymers are environment friendly materials due to their synthesis that uses safe monomers. Moreover, the biopolymers-based nanomaterials are advantageous, due to their morphological manipulation constructed at nanoscale. Biopolymeric nanomaterials can be derived from bacteria, algae, plants, fungi and animals. Furthermore, biomolecules can also be employed to derive biomaterials such as starch, alginate, chitin, chitosan and dextran. Among them, chitosan, starch and cellulose are the most commonly used biomaterials derived from microorganisms and plants [38, 42, 43].

Particularly, chitosan, a polysaccharide, is the most commonly used biomaterial for photodegradation of dyes and to remove pollutants from wastewater, due to its unique compositional and physiological properties. Biopolymeric nanomaterials are synthesized by integrating metal atoms into biopolymers by intramolecular hydrogen bonding. For example, starch can be incorporated to metal oxides to form biopolymeric nanostructures. Silver (Ag) is the most commonly used metal, and it has been reported to incorporate starch in molecular binding to form biopolymeric nanomaterials by adsorption/impregnation. Biopolymer-containing biomaterials comprise of solid and colloidal particles of 10 nm to 1 µm dimensions [25, 44, 45]. The as-synthesized biopolymeric substances for photodegradation of dyes can be characterized by microscopic, spectroscopic and other analytical techniques, as shown in Figure 3.1.

3.3.1.2 Metal-incorporated biopolymeric nanomaterials

As mentioned above, the metals and metal oxides can be incorporated in nanomaterials to enhance the efficiency of degradation of dyes such as azo dyes and methylene blue. The metal incorporation affects the porosity, rigidity and chemistry of biomaterial. The basics of photodegradation involve transfer of electrons, coupled with oxidation and reduction reactions. The semi-conductance nature of biomaterials can also be altered by incorporating metals and nanoparticles [29, 37]. The transfer of electrons at interface decreases the rate of oxidation of incorporated metals. The augmentation of reactivity of biomaterials not only enhances the lifetime of electrons but also increases the reactivity. The small particle size affords the transfer of electrons, thereby increasing the surface plasmon resonance. The incorporated metals act as stabilizer during the synthesis of biopolymeric nanomaterials. In industries, to remove dyes, resins are most commonly used as metal-incorporated biomaterial [42, 46].

Other than incorporation of metal oxides, the metal nanostructures can be grafted on different biomaterials to enhance the robustness and stability of biomaterials and provide them with functionalities. The metal monomeric nanostructures are different in size, and their incorporation in biopolymers enhances the homogeneity and renders them stable [47, 48]. The integration of microbial cellulose fibers with gold nanoparticles have been reported for the photodegradation of methylene blue dye, in less than 4 h. Owing to their substrate specificity, gold nanoparticles can degrade different types of organic pollutants, including dyes. The lanthanum metal can also be used to incorporate due to their f-orbital electrons, which reacts with the functional groups of dyes to form different complexes. The incorporation of lanthanum metals in biomaterials has been reported in photodegradation with comparatively better adsorption, specifically for organic dyes, for example, azo dyes. The biocomposite of lanthanum metal and metal oxide incorporated with chitosan have been reported to degrade 90% methylene blue dye in less than 40 min [37, 46, 49].

3.3.1.3 Metal oxide-incorporated bionanomaterials

Bionanomaterials integrated with metal oxide are now extensively used for photodegradation of complex chemical structures, such as methylene blue and azo dyes. The incorporation of metal oxides not only improves the chemical properties of the biomaterials but also enhances the properties of nanomaterials. For example, ZnO is now one of the most effective nanomaterials for photodegradation of dyes, due to its semiconductance, durability and activity. ZnO incorporated on starch has been recently reported for its enhanced dielectric properties as compared to other metal oxides [42, 50]. Cellulose-based metal oxide nanomaterials have been reported to exhibit better thermal stability as compared to other metal oxides, and, therefore, are now extensively used for bioremediation of wastewater. The nano-TiO_2 grafted on chitosan–acrylic acid has been reported to remove malachite green at neutral pH, by visible light irradiation. The nano-Fe_3O_4 grafted on chitosan has been reported to degrade X-3B dye-containing wastewater due to magnetic and dipole–dipole interactions. Recently, TiO_2 has been widely studied to synthesize metal oxide-incorporated bionanomaterials due to advantageous surface and photocatalytic properties, when these are exposed to visible light. Cyclodextrin, an oligosaccharide derived from starch, can also be used as derivative of biomaterial and can be integrated into nanomaterials and TiO_2 [47, 51].

The ZnO/carbon black can be grafted on cellulose acetate support to remove some azo dyes, including Congo red and methylene blue, under visible light. The MnO_2/cellulose bionanomaterials can degrade about 90% indigo carmine in less than 25 min, under acidic pH. Moreover, the bionanomaterials can be easily recovered from reaction system for 10 cycles, without any definite change in removal efficiency of dyes. It can be described as the photons are involved in photodegradation of complex and reactive azo dyes within few minutes through environmentally benign and reusable process. Some biomaterials are also employed as enzymes to degrade azo dyes. Laccase is an enzyme used for bioremediation of environmental pollutants, and it exhibits better potential to remove the contaminants from wastewater. The metal oxides, such as Fe_3O_4/SiO_2 anchored on immobilized laccase, have shown full azo dye breakdown, specifically, Congo red MX-5B, in less than 20 min [19, 44, 52].

3.3.1.4 Metal sulfide-incorporated bionanomaterials

Metal sulfides are also incorporated on bionanomaterials to aid their crystallization, such as ZnS can be incorporated on chitosan nanoparticles for photocatalysis activity, and it has been widely used for photodegradation of reactive dyes in aquatic environment. Not only Zn, but Cd and Pb have also been reported to synthesize their corresponding sulfide nanoparticles. For example, Klebsiella pneumonia has been reported to synthesize an electron-dense CdS nanostructure, based on its membrane for the photoreduction of reactive azo dyes like methyl orange through subsequent electron transfer reactions. Bionanomaterial based on nano-chitosan has also been used to degrade dyes by forming chitosan-based quantum dots. Also, the nano-photocatalyst has been integrated with ZnS to remove methylene blue dye under visible light. The metal sulfides improve the conductance and behave as semiconductors, thereby improving the overall efficiency of biomaterial. The ZnS/chitosan-based bionanomaterial has been found to induce 90 percent photodegradation of methylene blue dye through oxidation reactions under UV and visible irradiation in 1 h [13, 32, 53, 54].

3.3.2 Other bionanomaterials

Owing to the diverse compositional and chemical manipulations added by metal oxides and metal sulfides to biomaterials, these bionanomaterials are of special interest for researchers to clean the environment, due to their semiconductance and improved rate of generation of electron–hole pairs under visible light, which, thereby, enhance the photon activity, leading to the escalation of photocatalysis. The thin films of dual metal atoms, for example, Ag–TiO_2, can be integrated in resin biopolymers to enhance the photodegradation of rhodamine B dye, compared to pure metal oxides. These integrated bionanomaterials decrease the rate of recombination of electron–hole pairs. Also, a hybrid of TiO_2/Ag nanomaterials supported on carboxymethyl cellulase and gelatin biomaterials has been reported for enhanced [44, 52] removal activity of organic pollutants, due to their functional groups such as benzene and NH_3. Many other biochemicals can be incorporated on biopolymers to make photocatalytic bionanomaterials. Different coupled reactions have also expanded the photocatalytic activity, such as single metal atom integrated on nanomaterials, metal oxides, metal sulfides and dual metal systems incorporated on biomaterials to make integrated biopolymeric systems [11, 42]. The dual metal atoms can be beneficial due to their various band positions under sunlight exposure; this stops electron-hole pairs from recombining, thereby improving the overall photocatalytic activity of biomaterials. The synthesis of bionanocomposites has opened an avenue in the field of environmental sciences. Researchers are keen to prepare bionanocomposites based on single metal atoms, metal oxides and biopolymers to enhance the photodegradation of synthetic and reactive azo dyes [29, 30, 33].

The chitosan-based ZnO/AgCl dual metal atom bionanocomposites have been used as biopolymer supported on hydrogel beads for complete photodegradation of methylene blue dyes. The integration of chitosan as bionanocomposite has significantly increased the biodegradation, as a consequence of adsorption of organic pollutants in an eco-friendly manner. The chitosan can also be coated on other metals to oxidize the reactive dyes. For example, recently, chitosan has been coated on Cu_2SO_4 bionanocomposite and it was synthesized from *Psidium guajava* leaf extracts. It was found to oxidize the dyes and improved the removal of Congo red and methylene blue dyes, within 2 h. The rate of generation of electron–hole pair was enhanced due to the exposure of sunlight. The cellulose nanocrystals are under research for bleaching the bagasse pulp, by reacting it with ZnO/CuO biopolymers with dual metal oxide bionanocomposites. The as-synthesized bionanocomposites can be used to completely degrade rose bengal in 40 min. The immobilized enzymes such as laccase can be conjugated with chitosan based on Fe_3O_4 hybrids with enhanced magnetic properties, to remove reactive and synthetic blue and acidic blue dyes in less than 1 h [14, 32, 46, 53, 54].

3.3.3 Nanocomposites of biomaterials

Nanocomposites can be combined with different biomaterials with particle size range 1–100 nm. The use of nanocomposite biomaterials has gained much importance due to their improved physicochemical properties. These biomaterials are highly efficient for diverse functional and compositional applications in advanced science and technology, including mechanical strength, robustness, crystallinity, hierarchical porosity, toughness, electrical and thermal conductance, high mobility and durability. Recently, they have been used for bioremediation of different inorganic and organic pollutants by photocatalytic degradation. They can also be used for photodegradation and removal of organic and reactive dyes. The single/dual metal oxide-based bionanocomposites with impressive surface chemistry, biostability and tailored structural features make them highly feasible in the photodegradation of pollutants, mainly dyes. Owing to their remarkable photodegradation efficiency, tailored bandgap and efficient synthetic modulations, these bionanomaterials are now potentially employed for photodegradation of dyes. Certain metal oxides containing biomaterials can be supported on supports such as graphene oxide and biopolymer for successful photodegradation of organic dyes from aqueous environment [11, 28, 43, 50].

3.3.4 Green synthesis of bionanomaterials

Recently, researchers have focused their attention on synthesis of bionanomaterials based on green chemistry pathways. The green synthesis of biomaterials involves the use of biological materials, including plants, bacteria and algae carbohydrates and oligomers. Certain researches have shown green synthesis of bionanomaterials as an emerging frontier for better morphological manipulation with improved activity, as shown in Table 3.1. Therefore, these processes afford various benefits such as eco-friendly, simple, facile, conventional and economic pathways. Up to now, the metal oxide nanoparticles, for example, titanium dioxide NPs, zinc oxide NPs and iron oxide NPs has been synthesized for their use in photodegradation of organic pollutants such as dyes. Advanced oxidation processes (AOPs) have emerged as one of the most useful processes to decontaminate the environment from hazardous pollutants via the synthesis of ROS, for example, hydroxyl radicals (·OH) and superoxide free radicals ($\cdot O^{2-}$), and production of free radicals can further proceed to various pathways. It can follow photolysis, ozonation, Fenton reactions and photocatalysis [21, 48].

Photolysis and ozonation involve the use of peroxides along with UV light and ozone, respectively, by producing free radical. Fenton reactions are most currently being employed to treat wastewater, and it involves the use of peroxides along with ferrous ions as photocatalysts. Photocatalysis also involves the generation of free radicals using light absorbing materials, including semiconductors. Among all these reactions, photocatalysis is one of the most preferred methods, due to the photocatalysts, which use renewable energy and do not involve any hazardous chemical. Therefore, photocatalytic activity is now considered as the most effective green and sustainable approach to degrade dyes and other organic pollutants. The mechanism for biodegradation of pollutants and dyes can take two pathways, that is, indirect and direct pathways, of which indirect pathway is better. In the indirect pathway, the irradiation induces photoexcitation, and the semiconductor gets activated [16, 17, 37].

Table 3.1: Green synthesis of biomaterials for photodegradation of dyes.

Biomaterial	Morphology	Dyes	Photodegradation	Light	Reference
Aloe vera leaf extracts	Flake-like	Malachite green dye	58% (50 min)	Visible, UV	[55]
Mangrove extracts	Deformed and spherical	Blue 19 dye and Red 76 dye	~15% (120 min) ~15% (120 min)	Visible	[54]
Monsonia burkeana extracts	Spherical	Methylene blue dye	86% (180 min)	UV	[53]
Lagenaria siceraria extracts	Irregular spherical	Reactive green 19 dye	99% (60 min)	UV	[56]
Citrus aurantium extract	Spherical	Methylene blue dye	91% (150 min)	UV	[51]
Deinbollia pinnata extract	Semispherical	Methyl orange dye	99% (150 min)	UV	[57]
Acacia catechu extract	Spherical and hexagonal	Rhodamine B, rose, Bengal dye	99% (120 min)	Visible	[52]
Hydnocarpus alpina extract	Spherical	Methylene blue dye	96% (30 min)	UV	[58]
Longan seed extract	Hexagonal	Methylene blue, malachite green, methyl orange, orange II dyes	100% (180 min)	Visible	[50]

3.3.5 Cotton wastes as biomaterials to remove dyes

The cotton wastes in textile industries have been reported for their use in adsorption of reactive dyes in an economically viable pathway and at a large scale, due to ease of availability and processing. Cationic dyes are mainly removed by using cotton wastes from aqueous solutions. The wastewater containing dyes can be cleaned by different reaction mechanisms, such as Fenton reaction, oxidation, advanced oxidation pathway and ozonation. The photocatalytic adsorption of dyes from wastewater can be removed by using TiO_2 as a photocatalyst, due to its improved photocatalytic activity and stability. But the photocatalytic adsorption is enhanced under the radiations of UV light. The efficiency of cotton wastes for photodegradation of dyes can be enhanced by adsorption pathway due to its cost-effectiveness and high removal efficiency in low time of activity. The adsorption capacity of processed cotton waste such as polyacrylethyl trimethyl ammonium chloride cotton has been reported for its use to remove reactive scarlet 3BS dyes. An increase in concentration of processed cotton and temperature of the reaction was found to enhance its activity as adsorbent, which highlighted its efficiency in bioremediation of dyes [22, 59].

The maximum adsorption capacity can be obtained by Langmuir isotherm model by processed cotton, which showed its higher activity than other activated carbon materials and unmodified cotton. The significance of the effect of chemisorption and low uptake of BR2 dye and BR3 dye as substrates may occur due to unavailability of active sites and pores on the catalyst particles of processed cotton. The maximum sorption capacity can also be accomplished by polypyrrole-coated cotton fabrics to efficiently uptake methylene blue dye. The efficiency of cotton wastes can be improved by processing the waste for removal of dyes from aqueous solution, as shown in Table 3.2.

Table 3.2: Photodegradation of certain dyes in textile industry (adapted from [59]).

Precursors of cotton wastes	Activity
Pure cotton waste	Photodegradation of methylene blue dye
Impure cotton waste fabrics	Photodegradation of alizarin dye from aqueous solution
Processed cotton waste	Photodegradation of methylene blue and azo dyes
Cotton fiber	Methylene blue dye and rhodamine B 6 G dye removal
Waste of cotton seeds	Basic red 2, basic violet 3 dye removal
CNC from cotton seeds	Methylene blue dye removal from solution under light irradiation
Cotton hull wastes	Reactive blue dyes and red hues removal under visible light

3.4 Conclusion

The biodegradation of reactive and synthetic dyes by using photocatalysts is one of the most sustainable techniques in bioremediation. Here, in this chapter, we described some biomaterials and their derivatives supported on metals and metal oxides, by incorporating biopolymers to degrade and remove dyes. Certain biopolymers, including chitosan, gallic acid and cellulose afford a strong backbone to incorporate metal, metal oxides, metal sulfides, biopolymer, biohybrids and bionanocomposites due to their unique functional groups to facilitate the degradation of organic pollutants. In previous paragraphs, we explained the frequent use of novel and recently synthesized biomaterials and their composites for degradation of dyes, as compared to their pure counterparts. Additionally, the enhanced photocatalytic activity and reusability endorse their commercial use as photocatalyst. As a future prospect, commercial use of photocatalysts will not only aid in easy removal of dye effluents from textile industry at a large scale, but also facilitate the easy cleaning of harmful organic pollutants that are usually present in aquatics and water bodies. In this way, the threats to water bodies can be eliminated, while ensuring a promise to a safer future.

References

[1] Crini, G., E. Lichtfouse, G. Chanet and N. Morin-Crini. Applications of hemp in textiles, paper industry, insulation and building materials, horticulture, animal nutrition, food and beverages, nutraceuticals, cosmetics and hygiene, medicine, agrochemistry, energy production and environment: A review. Environmental Chemistry Letters, 2020. **18**(5): p. 1451–1476.

[2] Bhardwaj, D. and N. Bharadvaja, Phycoremediation of effluents containing dyes and its prospects for value-added products: A review of opportunities. Journal of Water Process Engineering, 2021. **41**: p. 102080.

[3] Gürses, A., M. Açıkyıldız, K. Güneş and M.S. Gürses. Dyes and pigments. 2016. Springer.

[4] Ahmad, I., et al., Lanthanum-zinc binary oxide nanocomposite with promising heterogeneous catalysis performance for the active conversion of 4-nitrophenol into 4-aminophenol. Coatings, 2021. **11**(5): p. 537.

[5] Ahmad, A., et al., A novel study on synthesis of egg shell based activated carbon for degradation of methylene blue via photocatalysis. Arabian Journal of Chemistry, 2020. **13**(12): p. 8717–8722.

[6] Bibi, S., et al., Photocatalytic degradation of malachite green and methylene blue over reduced graphene oxide (rGO) based metal oxides (rGO-Fe3O4/TiO2) nanocomposite under UV-visible light irradiation. Journal of Environmental Chemical Engineering, 2021. **9**(4): p. 105580.

[7] Kumar, A., G. Sharma, M. Naushad, H. Ala'a, A. García-Peñas, G.T. Mola, C. Si and F.J. Stadler, Bio-inspired and biomaterials-based hybrid photocatalysts for environmental detoxification: A review. Chemical Engineering Journal, 2020. **382**: p. 122937.

[8] Zhao, Y., Y. Wang, G. Xiao and H. Su. Fabrication of biomaterial/TiO2 composite photocatalysts for the selective removal of trace environmental pollutants. Chinese Journal of Chemical Engineering, 2019. **27**(6): p. 1416–1428.

[9] Hitam, C.N.C. and A.A. Jalil, Recent advances on nanocellulose biomaterials for environmental health photoremediation: An overview. Environmental Research, 2022. **204**: p. 111964.

[10] Xiao, K., T. Wang, M. Sun, A. Hanif, Q. Gu, B. Tian, Z. Jiang, B. Wang, H. Sun, J. Shang and P.K. Wong. Photocatalytic bacterial inactivation by a rape pollen-MoS2 biohybrid catalyst: synergetic effects and inactivation mechanisms. Environmental Science Technology, 2019. **54**(1): p. 537–549.

[11] Li, L., Z. Xu and X. Huang. Whole-Cell-Based Photosynthetic Biohybrid Systems for Energy and Environmental Applications. ChemPlusChem, 2021. **86**(7): p. 1021–1036.

[12] Ahmad, A., et al., *A novel study on synthesis of egg shell based activated carbon for degradation of methylene blue via photocatalysis*. Arabian Journal of Chemistry, 2020. **13**(12): p. 8717–8722.

[13] Hu, P., X. Hu, C. Chen, D. Hou and Y. Huang. Biomaterial-assisted synthesis of AgCl@ Ag concave cubes with efficient visible-light-driven photocatalytic activity. CrystEngComm, 2014. **16**(4): p. 649–653.

[14] Matmin, J., I. Affendi, S.I. Ibrahim and S. Endud. Additive-free rice starch-assisted synthesis of spherical nanostructured hematite for degradation of dye contaminant. Nanomaterials, 2018. **8**(9): p. 702.
[15] Gao, G., Y.W. Jiang, H.R. Jia and F.G. Wu, Near-infrared light-controllable ondemand antibiotics release using thermo-sensitive hydrogel-based drug reservoir for combating bacterial infection. Biomaterials, 2019. **188**: p. 83–95.
[16] Hasan, G.G., H.A. Mohammed, M. Althamthami, A. Khelef, S.E. Laouini and S. Meneceur, Synergistic effect of novel biosynthesis SnO2@ Fe3O4 nanocomposite: A comprehensive study of its photocatalytic of Dyes & antibiotics, antibacterial, and antimutagenic activities. Journal of Photochemistry and Photobiology A: Chemistry, 2023. **443**: p. 114874.
[17] Abbasnia, A., A. Zarei, M. Yegane, H.R. Sobhi, M. Gholami and A. Esrafili. Removal of tetracycline antibiotics by adsorption and photocatalytic-degradation processes in aqueous solutions using metal organic frameworks (MOFs): A systematic review. Inorganic Chemistry Communications, 2022: p. 109959
[18] DeLaive, P.J., J.T. Lee, H.W. Sprintschnik, H. Abruna, T.J. Meyer and D.G. Whitten. Photoinduced redox reactions of hydrophobic ruthenium (II) complexes. Journal of the American Chemical Society, 1977. **99**(21): p. 7094–7097.
[19] Kalyanasundaram, K. and M. Grätzel. Light induced redox reactions of water soluble porphyrins, sensitization of hydrogen generation from water by zincporphyrin derivatives. Helvetica Chimica Acta, 1980. **63**(2): p. 478–485.
[20] Salehi, G., R. Abazari, and A.R. Mahjoub, *Visible-light-induced graphitic–C3N4@ nickel–aluminum layered double hydroxide nanocomposites with enhanced photocatalytic activity for removal of dyes in water.* Inorganic Chemistry, 2018. **57**(14): p. 8681–8691.
[21] Chen, Y., et al., *Green synthesis of Ag–TiO2 supported on porous glass with enhanced photocatalytic performance for oxidative desulfurization and removal of dyes under visible light.* ACS Sustainable Chemistry & Engineering, 2018. **6**(10): p. 13276–13286.
[22] Emam, H.E., et al., *Recyclable photocatalyst composites based on Ag3VO4 and Ag2WO4@ MOF@ cotton for effective discoloration of dye in visible light.* Cellulose, 2020. **27**(12): p. 7139–7155.
[23] Agarwal, R. and A.J. García, *Surface modification of biomaterials*, in *Principles of regenerative medicine.* 2019, Elsevier. p. 651–660.
[24] Aghajani, M. and F. Esmaeili, *Anti-biofouling assembly strategies for protein & cell repellent surfaces: a mini-review.* Journal of Biomaterials Science, Polymer Edition, 2021(just-accepted): p. 1–21.
[25] Dutta, S., et al., *Bio-derived metal and metal oxide incorporated biopolymer nanocomposites for dye degradation applications: A review.* Octa Journal of Biosciences, 2021. **9**(1)
[26] Muthukumar, H., et al., *Biogenic synthesis of nano-biomaterial for toxic naphthalene photocatalytic degradation optimization and kinetics studies.* International Biodeterioration & Biodegradation, 2017. **119**: p. 587–594.
[27] Kumar, A., et al., *Bio-inspired and biomaterials-based hybrid photocatalysts for environmental detoxification: A review.* Chemical Engineering Journal, 2020. **382**: p. 122937.
[28] Mondal, S., *Methods of dye removal from dye house effluent – an overview.* Environmental Engineering Science, 2008. **25**(3): p. 383–396.
[29] Dassanayake, R.S., S. Acharya, and N. Abidi, *Recent advances in biopolymer-based dye removal technologies.* Molecules, 2021. **26**(15): p. 4697.
[30] Hitam, C. and A. Jalil, *Recent advances on nanocellulose biomaterials for environmental health photoremediation: An overview.* Environmental Research, 2022. **204**: p. 111964.
[31] Vinay, S., et al., *Hydrothermal synthesis of gold nanoparticles using spider Cobweb as novel biomaterial: application to photocatalytic.* Chemical Physics Letters, 2020. **748**: p. 137402.

[32] Al-Kahtani, A.A., *Photocatalytic degradation of rhodamine B dye in wastewater using gelatin/CuS/PVA nanocomposites under solar light irradiation*. Journal of Biomaterials and Nanobiotechnology, 2017. **8**(01): p. 66.

[33] Rajasekar, M., *Recent trends in rhodamine derivatives as fluorescent probes for biomaterial applications*. Journal of Molecular Structure, 2021. **1235**: p. 130232.

[34] Darvishi Cheshmeh Soltani, R., A. Rezaee, and A. Khataee, *Combination of carbon black–ZnO/UV process with an electrochemical process equipped with a carbon black–PTFE-coated gas-diffusion cathode for removal of a textile dye*. Industrial & Engineering Chemistry Research, 2013. **52**(39): p. 14133–14142.

[35] Saeed, M., et al., *Ag@ Mn_xO_y: an effective catalyst for photo-degradation of rhodamine B dye*. Environmental Chemistry Letters, 2018. **16**(1): p. 287–294.

[36] Rajeswari, A., S. Vismaiya, and A. Pius, *Preparation, characterization of nano ZnO-blended cellulose acetate-polyurethane membrane for photocatalytic degradation of dyes from water*. Chemical Engineering Journal, 2017. **313**: p. 928–937.

[37] Sirajudheen, P., et al., *Technological advancement in photocatalytic degradation of dyes using metal-doped biopolymeric composites – present and future perspectives*, in *Nanomaterials and Nanocomposites for Environmental Remediation*. 2021, Springer. p. 205–255.

[38] Devanesan, A.A. and D.J. Ali, *Nanobiocatalysts and photocatalyst in dye degradation*. Physical Sciences Reviews, 2021.

[39] Wang, Q., et al., *Photodegradation of textile dye Rhodamine B over a novel biopolymer–metal complex wool-Pd/CdS photocatalysts under visible light irradiation*. Journal of Photochemistry and Photobiology B: Biology, 2013. **126**: p. 47–54.

[40] Nouri, L., et al., *Elaboration and characterization of photobiocomposite beads, based on titanium (IV) oxide and sodium alginate biopolymer, for basic blue 41 adsorption/photocatalytic degradation*. International Journal of Biological Macromolecules, 2020. **151**: p. 66–84.

[41] Manimohan, M., et al., *Synthesis and characterisation of novel Cu (II)-anchored biopolymer complexes as reusable materials for the photocatalytic degradation of methylene blue*. RSC Advances, 2020. **10**(31): p. 18259–18279.

[42] Mallakpour, S., F. Sirous, and C.M. Hussain, *A journey to the world of fascinating ZnO nanocomposites made of chitosan, starch, cellulose, and other biopolymers: Progress in recent achievements in eco-friendly food packaging, biomedical, and water remediation technologies*. International Journal of Biological Macromolecules, 2021. **170**: p. 701–716.

[43] Ribeiro, J.S. and C.M. Veloso, *Microencapsulation of natural dyes with biopolymers for application in food: A review*. Food Hydrocolloids, 2021. **112**: p. 106374.

[44] Khalil, A., et al., *Synthesis and catalytic evaluation of silver@ nickel oxide and alginate biopolymer nanocomposite hydrogel beads*. Cellulose, 2021. **28**(18): p. 11299–11313.

[45] Eswaran, S.G., H. Narayan, and N. Vasimalai, *Reductive photocatalytic degradation of toxic aniline blue dye using green synthesized banyan aerial root extract derived silver nanoparticles*. Biocatalysis and Agricultural Biotechnology, 2021. **36**: p. 102140.

[46] Vigneshwaran, S., et al., *Photocatalytic performance of chitosan tethered magnetic Fe2O3-like (3D/2D) hybrid for the dynamic removal of anionic dyes: Degradation and mechanistic pathways*. International Journal of Biological Macromolecules, 2021.

[47] Vigneshwaran, S., et al., *In situ fabrication of ternary TiO2 doped grafted chitosan/hydroxyapatite nanocomposite with improved catalytic performance for the removal of organic dyes: Experimental and systemic studies*. Colloids and Surfaces A: Physicochemical and Engineering Aspects, 2021. **611**: p. 125789.

[48] Sarkar, S., et al., *Green polymeric nanomaterials for the photocatalytic degradation of dyes: a review*. Environmental Chemistry Letters, 2020: p. 1–12.
[49] Sirajudheen, P. and S. Meenakshi, *Facile synthesis of chitosan-La3+-graphite composite and its influence in photocatalytic degradation of methylene blue*. International Journal of Biological Macromolecules, 2019. **133**: p. 253–261.
[50] Chankaew, C., et al., *Microwave synthesis of ZnO nanoparticles using longan seeds biowaste and their efficiencies in photocatalytic decolorization of organic dyes*. Environmental Science and Pollution Research, 2019. **26**(17): p. 17548–17554.
[51] Mulugeta, H., *Synthesis of titanium dioxide/reduced graphene oxide nanocomposite using peel extracts of Citrus aurantium and Musa acuminata for photocatalytic degradation of methylene blue from wastewater*. 2020, ASTU.
[52] Kaur, A., et al., *Sustainable preparation of Fe (OH) 3 and α-Fe2O3 nanoparticles employing Acacia catechu extract for efficient removal of chromium (VI) from aqueous solution*. Environmental Nanotechnology, Monitoring & Management, 2021. **16**: p. 100593.
[53] Makofane, A., D.E. Motaung, and N.C. Hintsho-Mbita, *Photocatalytic degradation of methylene blue and sulfisoxazole from water using biosynthesized zinc ferrite nanoparticles*. Ceramics International, 2021.
[54] Nathan, V.K., P. Ammini, and J. Vijayan, *Photocatalytic degradation of synthetic dyes using iron (III) oxide nanoparticles (Fe_2O_3-Nps) synthesised using Rhizophora mucronata Lam*. IET Nanobiotechnology, 2019. **13**(2): p. 120–123.
[55] Seifunnisha, O. and J. Shanthi, *Influence of Aloe vera and PEG on the evaluation of photocatalytic degradation of MG dye under UV light and visible light irradiation of ZnO nanomaterials*. Optik, 2021. **248**: p. 168064.
[56] Kalpana, V. and V.D. Rajeswari, *Synthesis of palladium nanoparticles via a green route using Lagenaria siceraria: assessment of their innate antidandruff, insecticidal and degradation activities*. Materials Research Express, 2018. **5**(11): p. 115406.
[57] Rufai, Y., S. Chandren, and N. Basar, *Influence of solvents' polarity on the physicochemical properties and photocatalytic activity of titania synthesized using deinbollia pinnata leaves*. Frontiers in Chemistry, 2020. **8**: p. 1144.
[58] Ganesh, M., et al., *Hydnocarpus alpina Wt extract mediated green synthesis of ZnO nanoparticle and screening of its anti-microbial, free radical scavenging, and photocatalytic activity*. Biocatalysis and Agricultural Biotechnology, 2019. **19**: p. 101129.
[59] Rizal, S., et al., *Cotton wastes functionalized biomaterials from micro to nano: a cleaner approach for a sustainable environmental application*. Polymers, 2021. **13**(7): p. 1006.

Sadaf Tariq, Awais Ahmad, Rafael Luque, Mabkhoot Alsaiari, Farid A. Harraz

4 Metals removal from wastewater via biomaterials

Abstract: Potable and safe water is required for all living organisms to thrive. Access to water has been ranked at the top among the requisites to define standard of life. There is a continuous decline in water quality due to anthropogenic and industrial activities, and poor water quality severely affects water resources, causing disasters. Globally, many initiatives have been taken to clean water. Researchers are now focusing on wastewater bioremediation protocols to decontaminate water from organic and inorganic pollutants. Water fortification requires the use of natural materials instead of non-biodegradable chemicals. Bioderived materials can be used to detoxify the high reactivity of metals in wastewater, considering selectivity and environmental benignity, over conventional technologies. Biomass-derived biomaterials are now used as alternatives for biosorption of heavy metals from wastewater as an innovative technology. The biotechnological exploitation of biomaterials depends upon the reusability of biosorption materials. This chapter specifically aims to provide information about the cheaper biomaterials that occur as natural adsorbents and biosorbents, to clean up heavy metals from wastewater. The metal binding properties and mechanisms of biosorbents are also discussed to elaborate the use of biomaterials for bioremediation and environment cleaning, along with their future prospects.

Keywords: Biomaterials, wastewater, bioremediation, metals removal, biosorption

4.1 Introduction

In recent decades, one of the most pressing environmental concerns is the availability of potable and safe water to living organisms, as water pollution is one of the important worldwide issues. The presence of inorganic contaminants in water, including heavy metals, is due to human or geological activities. It presents a serious threat to all living organisms of aquatic and terrestrial ecosystems [1, 2]. Annually, many tons of inorganic pollutants are discharged from industries into aquatic ecosystems, which change the physicochemical properties of natural water, making it harmful for use. For example, the presence of inorganic pollutants may decrease the penetration and absorption of sunlight, which halts photosynthesis in contaminated aquatic ecosystems [3–6].

https://doi.org/10.1515/9783110768749-004

Therefore, it is high time to search and explore some efficient materials and technologies to degrade these contaminants from wastewater. Up to now, different treatment technologies have been reported for environmental cleaning, by either removing or degrading inorganic and toxic contaminants, specifically metals and metal ions. These strategies include advanced oxidation processes (AOPs), biological/electrochemical oxidation, photodegradation, photocatalysis, biodegradation and chlorination [7–9]. Certain traditional strategies are also in common use to decontaminate water, such as membrane filtration, coagulation/flocculation and purification, etc.

These strategies are not sufficient to clean up water completely due to certain drawbacks such as being laborious, cost-ineffectiveness, take a lot of time, and generate hazardous byproducts and undesired reaction intermediates, preventing their use on a large scale [10–13]. Moreover, some chemicals and reaction intermediates that are synthesized during the reaction are resistant to biodegradation and photocatalysis. Up to now, the use of natural materials for bioremediation of wastewater to degrade metals by using microorganisms such as algae and bacteria is somewhat typical [14–17]. These limitations can be overcome by using different water remediation techniques that would be more suitable than conventional strategies, such as how adsorption is now considered as one of the most appealing strategies.

Biomaterials can be used as natural biosorbents due to their easy-handling, time saving ability and environment effective methods for degrading inorganic pollutants. Recently, various materials have been reported as bioadsorbents, such as biopolymeric matrices, agroindustrial wastes, biomass, chitosan, cellulose, cotton wastes, inorganic matrices and hybrid nanoporous solids (MOFs) to degrade metals in wastewater [18–21]. This chapter aims to provide a brief description on the current research activities in the field of biomaterials by evaluating whether biosorption can eventually be used as an efficient alternate approach over other conventional approaches in this field. We discuss that biomaterials are effective in removing metals by biological treatment, as metals tend to accumulate in biological environments. This perspective highlights the most important developments in wastewater treatment technologies. We offer a short note on the future challenges for in-depth study. Additionally, current challenges in bioremediation approaches by biomaterials are presented.

4.2 Metals removal from wastewater

Metals can be removed from wastewater by using biomaterials; it is an innovative approach for environment cleaning. Overall, bioremediation has emerged as one of the most sustainable approaches for biological cleaning of the environment from various pollutants. In this regard, biomaterials catalyze the removal/degradation of pollutants by using biological systems. Biomaterials are therefore important as they can be used to clean water streams, acting as low-cost biosorbents. Different cost-effective and biodegradable biomaterials, such as seaweeds, yeasts, bacteria, molds and agricultural products, can be used to remove metal ions from wastewater. However, it is necessary to find biomaterials for industrial and large scale applications for water treatment purposes [8, 10, 22].

During water treatment, particulate suspensions can be removed by flocculants and coagulants –neutralize the charged suspensions by coagulation or aggregate the suspension by flocculation. Recently, nanomaterials-based biomaterials have been the subject of study for their efficient application as coagulants and flocculants. Municipal wastewater can be cleaned using anionized nanomaterials. The functionalized biomaterials exhibited efficiency in coagulation and flocculation water treatment [4, 23, 24].

Many agricultural products, which otherwise are of no use, can be used as highly valued biomaterials for cleaning water – bioremediation of metallic contaminants – as shown in Table 4.1. Adsorption isotherms reveal that biomaterials show maximum activity at a definite range of pH. Non-competitive and concentration-dependent isotherms can be studied to model and predict the concentration of metals and biomaterials. Furthermore, the capacity of biosorption of certain plants has been studied to remove As, Cr, Cu, Zn, Ni, Fe, Hg, Mn and Pb from wastewater. Pb metal ions (with concentration of up to 2 g), along with Zn and Cu, can be removed by using plants as biomass. Phosphate-treated rice husk, at different concentrations, can be used as biosorbents to remove Cr and Zn. *Bacillus* spp. have been reported for their high metal removal efficiency – removing Pb, Cu, Co and Zn from wastewater by biosorption. *Micrococcus* spp. are also very efficient in removing metals from wastewater. *Mucor* spp. are fungi remarkable for its metal removal efficiency from wastewater by biosorption. *Amanita* spp. can also treat water by biosorption, offering different adsorption efficiencies at different concentrations with [14, 15, 22, 25, 26].

Bacteria can be used to remove metals – *Ecklonia* spp. has been reported to remove Cr metal ions at low pH. Jack fruit extracts can be treated with acid to improve their bioadsorption as a biomass. Complete removal of some metals has been reported by treating them with 0.001 M HCl. By using apricot seeds, biosorption can be achieved at acidic pH. These biomaterials can be reused with retained activity by successive adsorption-desorption cycles. *M. oleifera* seeds have also been reported to remove metals from wastewater. Rice husk can be used to remove Zn, As and Hg, with different adsorption capacities at different concentrations. Date pits can be used to remove Cu and Cd from wastewater [24, 39]. Pine sawdust has been reported for its efficient Zn removal efficiency. Coir pith has also been reported to remove hexavalent Cr from water by biosorption. Sugarcane bagasse is an agro-based by-product; it has been reported to remove metals from wastewater by reducing the

Table 4.1: Biomaterials reported for biosorption of metals from wastewater.

Biomaterial	Nature	Metals removed by biomaterial	Capacity of metals removal (mg/g)	Reference
Cystoseira indica	Algae	Ur(VI)	318	[27]
Anabaena sphaerica	Algae	Cd(II)	111	[28]
Maugeotia genuflexa	Algae	As(III)	53	[27]
Cladophora hutchinsiae	Algae	Se(IV)	74	[28]
Trametes versicolor	Fungi	Cu(II)	140	[29]
Mucor rouxii	Fungi	Zn(II)	53	[30]
Gliomastix murorum	Fungi	Cu(II)	9	[29]
Trametes versicolor	Fungi	Pb(II)	280	[31]
Corynebacterium glutamicum	Bacteria	Ni(II)	111	[27]
Spirulina platensis	Bacteria	Hg(II)	428	[27]
Geobacillus toebii	Bacteria	Cu(II)	41	[32]
Banana peel	Plant	Pd(II), Cd(II)	2, 5	[33]
Lemon peel	Plant	Ni(II), Cu(II), Cd(II)	22, 50	[34]
Orange peel	Plant	Pb(II), Cd(II), Zn(II), Cr, Cu	476, 293, 118	[35]
Almond green hull	Agro-by-product	Co(II)	45	[36]
Sugarcane bagasse	Agro-by-product	Zn(II), Cu(II)	45, 9	[37]
Corncob	Agro-by-product	Zn(II)	49	[38]

metals [34, 37]. Figure 4.1 illustrates the living and non-living sources of biomaterials to remove metals from wastewater.

Bioaccumulation is an alternative to biosorption, and can be used for the removal of metals from wastewater. Both biosorption and bioaccumulation are nature-driven processes, which mainly depend upon pH and temperature. The only difference between the two processes is that bioaccumulation involves living cells while biosorption involves non-living cells. The biological cells take up metals and other contaminants by involving active metabolic activities, which happens through energy consumption. Biological cells survive only in a definite environment under defined conditions; therefore, even at low concentrations, bioaccumulation is more efficient than biosorption. But living cells should be handled with care, and thus are avoided. But due to its low-cost and biodegradable nature, biosorption is preferred over bioaccumulation [23, 40–42].

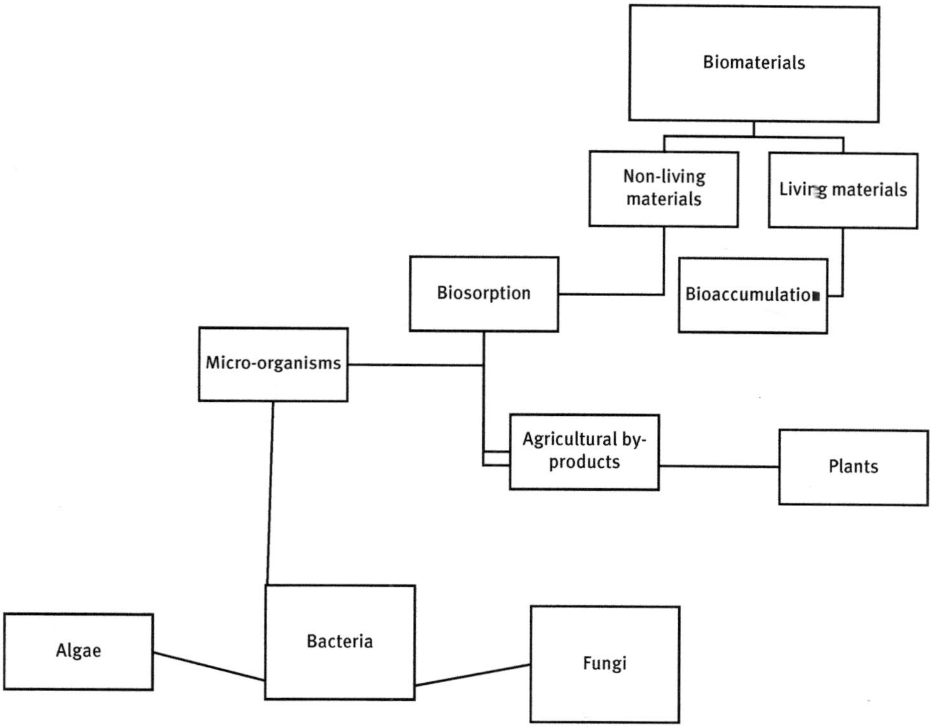

Figure 4.1: An illustration of different biomaterial sources for metal removal from wastewater.

As compared to bioaccumulation, the shortcomings of biosorption are fewer, such as smaller size of biomaterial used, nonliving form and incorporation/separation from the system. Similarly, several factors, including active functional groups on the surface of biomaterial, reaction conditions, pH, ionic strength, efficiency of reaction and coexistence of ions, affect the targeted adsorption of metals from wastewater. The pH of the solution is essentially monitored during the reaction. The precise understanding of biosorption mechanism can be used to design specific biomaterials for specific biosorption of metals from wastewater [31, 33, 43, 44].

4.2.1 Nanomaterial-based adsorbents

The growing trend of nanochemistry urged researchers to develop nanomaterial-based bioadsorbents for water treatment. Extensive research has been carried out and reported to develop effective biomaterials for removal of metals from wastewater. The optical, magnetic, and thermal characteristics of biological materials and their catalytic properties can be tailored by tuning their structures and composition. Nanomaterials provide the biomateials with certain advanced properties such as enhanced adsorption, high binding capacities, good interfacial properties and a particularly high surface area at nanoscale, in comparison to other molecular counterparts. Additionally, the nanomaterial-based bioadsorbents can be finely and precisely tuned to modify their structures and properties. Owing to its simple nature, adsorption is one of the most significant approaches in water treatment. The benchmark for an efficient adsorbent includes cost, robustness, adsorption capacity, integrity, surface area, ease of reusability and environmental benignity [5, 23, 36, 45, 46].

Due to the limitations caused by conventional adsorbents, recent efforts are focused on developing sustainable and efficient bio-based alternatives for water treatment. Researchers have reported the use of nanocellulosics as nanomaterial-based bioadsorbents to remove metals from water. Nanocellulosics have been efficiently used to remove cadmium(II) ions from wastewater extracted from *spinifex. Spinifex* belongs to the genus Triodia, an abundant and not commonly used grass with a wide distribution across all continents. The nitro-oxidation approach can be employed to develop nanocellulosics with low crystallinity, high surface charge and remarkable hydrophilicity. The high efficacy of nanocellulosics was displayed by suspension, which was substantially better than conventional adsorbents. Therefore, spinifex has been reported to harness its potential for the bioremediation of water [17, 33, 47–49].

Nanomaterial-based aerogels have been recently reported to remove highly toxic and reactive mercury metal ions from polluted water due to their simple, reusable and environmental benign nature as bioadsorbents. This study demonstrated the use of surface-tailored nanomaterial-based biomaterials. Nanocellulosic aerogels can be obtained by the freeze drying synthesis approach. They can be extracted from bamboo. The TEMPO functional group on the bamboo-based nanocellulosics was oxidized into NFC in the presence of mercaptopropysiloxane. The thiol groups are adsorbed on the nanocellulosics surface, while the metal ions can be efficiently removed. Nanocellulosics thereby demonstrate the highly efficient nanomaterials in the field of biosorption [16, 30, 31, 41].

4.2.2 Biosorption to remove metals

Heavy metals can be removed by different biomaterials; it takes place by biosorption. Biosorption, by using biomaterials, involves metabolism-independent passive transport of a solid phase (sorbent) in a solvent (liquid phase) and helps to dissolve the sorbed species. Biomaterials are good sorbents and possess a high affinity for metals and metal ions [39, 44]. The interactions between metal ions and biomaterials involve adsorption, complexation and chemisorption on the surface of the sorbents, through ion exchange. Agricultural residues can be used to remove toxic metals from wastewater by adsorption, complexation and ion exchange through a facile, economic, conventional, efficient and potential technique [47, 50].

The proposed mechanism for biosorption involves the sorption of cationic metal species through ion exchange between the protons of metals and the counter ions in the biomass. The cationic ion exchange takes place between the acidic and basic functional groups, and the metal ions. The biomass is active and takes up cations from the metal ions in wastewater. Complexation is another mechanism of biosorption, followed by some vegetal biomaterials, such as lignin and cellulose. They possess specific functional groups, including alcoholic, ketonic and carboxylic groups, which participate in the complexation with metal ions [51, 52].

Owing to water treatment, protein can be used as biomaterials for the biosorption of metal ions from wastewater. Most preferably, keratin has been reported to be used as a bioadsorbent due to its biodegradable and biocompatible nature. Researchers have reported the use of a modified keratin for its potential application of water decontamination, as a renewable source. Most of the research is based on keratin extracted from chicken feathers-based biosorbents. Chicken feathers were modified in a work conducted by [45] to analyze its biosorption properties. The modified keratin removed Pb(II) at different concentrations, different pH levels and at different temperatures. Landin-Sandoval et al. [43] reported the biosorption of Zn(II) by using keratin as a biosorbent. Under acidic pH, the uptake of Zn was found to be maximum at 30 °C. Other sources of keratin protein to remove Cd, Ni, Cr, Zn, Ar, Ag, Pb and Cu from wastewater could be egg shell, wool and horn hoof [47, 48].

In acidic pH, the binding efficiency of a biomass and a heavy metal can be calculated by the degree of dissociation of the functional groups. It can be investigated to analyze the dissociation of metal ions during biosorption. The dissociation of some earth and alkaline earth metals has been studied to determine the ion exchange between metals and the biomass. The functional groups on biomaterials may involve the sorption of metal ions, and it can be characterized by using certain techniques. Many characterization technologies have been employed to determine the characteristic mechanism of biosorption by studying various factors and components [23, 25]. The oxidation state of metals, which preferably bind to biomaterials and the functional groups involved in the biosorption process, can be studied by Fourier-transformed infrared spectroscopy (FTIR).

Other useful techniques may include UV–Vis spectroscopy, atomic adsorption spectroscopy, scanning electron microscopy, transmission electron microscopy, X-ray diffraction analysis and X-ray photoelectron spectroscopy, as shown in Table 4.2. UV– Vis spectroscopy is very important as it is used to detect the oxidation state of metals and the information about the oxidation state of metals provides direct information about its energy to absorb the exciton and the dissociation energy [46, 53]. The SEM and the images captured with a TEM can reveal morphological data of the adsorbent, before and after biosorption. The atomic energy levels during biosorption can also be probed by using the photoelectron spectroscopy. The nature of metal ions and the difference in their morphology after biosorption can also be understood by XRD analysis. The XPS further confirms the oxidation states of the bound metal ions. In the most recent reported studies, the ion exchange process during biosorption between the metal ions and the biomaterials was confirmed by using the scanning electron microscopy–energy dispersive X-ray analysis (SEM–EDX) [24, 26, 40].

Table 4.2: Common characterization techniques for the biosorption of metal ions.

Analysis	Techniques	Applications
Spectroscopic analysis	X-ray diffractometry (XRD)	Nature and behavior of metal ions
	Energy dispersive x-ray spectroscopy (EDX)	Ion exchange between metal ions and biomaterials
	Atomic absorption spectroscopy (AAS)	Determination of concentrations of metal ions in wastewater
	X-ray photoelectron spectroscopy (XPS)	Oxidation state of heavy and light metal atoms
	UV-Visible spectroscopy	Absorption properties
	Fourier transform infrared spectroscopy (FTIR)	Oxidation state and functional groups determinations
	Photoelectron spectroscopy	Atomic molecular energy levels
Microscopic analysis	Scanning electron microscopy (SEM)	Structural features
	Transmission electron microscopy (TEM)	
	Atomic force microscopy (AFM)	
Other analysis	Dynamic light scattering (DLS)	Surface charges

4.2.3 Bioremediation to remove metals

The conventional approaches used to remove metal ions from wastewater are cost-ineffective and environment non-friendly. To avoid the environmental problems caused by wastewater disposal, biosorption can be used to remove the accumulated toxic metal ions by using biomass. The bioremediation technology explores biomaterials to form complexes with accumulated metal ions by using the available ligands and functional groups on the surface of biomaterials. Specifically, cell walls of microorganisms, algae, fungi and plants are responsible for the biodegradation of metal ions. This phenomenon can be commonly employed due to its cost effective nature to purify wastewater [50, 51]. The bioremediation technology is so powerful that drinking water quality can also be obtained from wastewater. The major advantages of biosorption through the use of biomaterials, over conventional approaches, may include: cost-effectiveness, high efficiency, minimal use of chemicals, reusability of biosorbents and metal atoms recovery [22].

Heavy metals and metal ions are toxic even at low concentrations when they exceed a certain threshold; they persist as water pollutants. Through activated carbon, the metals are precipitated on the surface through metal ion sequestration, while it is quite an expensive strategy to remove metal ions. A recent approach is the use of cotton waste to remove metals from wastewater in a cheaper way. The cotton waste acts as precursors in the adsorbent media. The sorption capacity of cotton waste to remove/degrade metal ions from wastewater is comparatively better than other conventional methods. Pb, Cr and Cd ions were easily removed by using waste cotton yarn [48, 52]. Waste cotton yarn has been reported to remove Pb, Cd, Cr and As metal ions from the wastewater of textile industries. Waste cotton fibers are also reported to remove inorganic pollutants including metals from aqueous solution at a high efficiency. Cotton hull has been reported to remove Cr(VI) from aqueous solutions and Pb(II) from gold mining liquid effluents, at high concentrations [52].

4.2.4 Emerging wastewater treatment methods

Advanced oxidation processes (AOPs) for wastewater purification have recently been reported as an emerging wastewater treatment technology. AOPs generate radicals and involve in-situ chemical oxidation reactions. Owing to the high oxidation nature, the generated radicals can mineralize the contaminants very effectively. Through AOPs, different non-biodegradable and highly stable aquatic contaminants can be removed very efficiently. Their mineralization, either decreases or completely removes the pollutants from wastewater. The chances of secondary pollution are also minimized due to the use of bio-inspired materials for the biodegradation of metal ions as no toxic reaction intermediates are generated [2, 7, 11, 49, 54, 55].

Photocatalysis is another promising method and has been recently emerging as one of the most sustainable methods for the decontamination of wastewater. It involves the use of renewable energy resources; semiconductors act as photocatalysts in this regard by acting as a preferable technology to remove the organic and inorganic pollutants from aquatic systems by using solar light and suitable photocatalysts. Hydroxide and superoxide radicals are generated on illumination through a photocatalyst in the suspension of aquatic pollutants, and play a vital role in the biodegradation of environmental pollution and aids in the sustainability of the environment. Owing to the efficiency of the emerging photocatalytic technology for wastewater treatment, researchers are much more focused to study the properties and factors affecting their remarkable activity. The role of semiconductors and metals as photocatalysts can be determined and illustrated by studying their properties. The support materials also aid the interactions between pollutants and help in the facile and subsequent transport of free radicals for the degradation of metals in wastewater [14, 25, 56].

An appropriate and ideal support can be provided by using biomaterials to enhance the strong connections between the photocatalysts, and to provide a large surface area, better adsorption and enhanced stability. Owing to the physio-chemical properties, biopolymers used as biomaterials can efficiently trap the photogenerated electrons for enhanced biodegradation. However, the decreased solubility, thermal stability, surface area and weak mechanical strength have restricted researchers from using these in industrial wastewater treatment. Correspondingly, to attain suitable photocatalysis, a considerable attention is focused on organic biopolymers, including chitin, chitosan, starch, hydroxyapatite, keratin and other biomolecules, as they are biodegradable and easily available. Furthermore, the porosity and abundance of the polar functional groups help in the enhanced interaction between metals and biopolymers [2, 46, 47].

4.2.5 Biosorption by microbial biomass

The microbial biomass binds the selective contaminants in wastewater by biosorption. Bioaccumulation is, conversely, based on the complex phenomenon of active metabolic processes. The biosorption of dead biomass is based on the strength of interaction between the biosorbent and the contaminants. The biomass can be used to remove metal ions, either directly or by subjecting it to modification. But to achieve better results, the biomass is subjected to chemical modification. The dead biomass is immobilized to improve the mechanical strength, enhance the separation of metals and ease the process of biosorption. The toxic and heavy metals can be removed by using biomass such as algae as biosorbents, with high selectivity and mechanical properties. The constituents of algal cell wall, such as brown algae, alginate and fucoidan, have been reported to chelate heavy metals. Some algal strains, including *Chlorella* spp., *Spirulina* spp., and *Cladophora* spp., have been reported to exhibit significant removal efficiency in an aqueous solution [25, 26, 29, 31].

Fungal strains as biomass have been reported to remove metals. Their ease of availability from fermentation industries has increased their use as biosorbents. Their cell wall constitutes chitin, mannin and proteins, which are made up of carboxyl, hydroxyl, phosphate, thiol, amide and imidazole functional groups. Their dead biomass possesses biosorption capacity to remove pollutants. Certain fungal species, including *Aspergillus niger, Schizophyllum commune, Rhizopus arrhizus* and *Mucor* spp. have been reported in the removal of metals from wastewater. Crucial environmental conditions such as pH, modification strategies, nature of fungi and concentration greatly affect the biosorption of metals. Similarly, the cell walls of bacterial species interact with the contaminants in wastewater, such as metal ions [30, 31, 41].

4.2.6 Low-cost biomaterials as bioadsorbents

Conventional approaches using conventional adsorbents, such as activated carbon, sludge, soya cake, rubber tires, fly ash and rice husk-based substrates have been reported to remove metal ions from wastewater. Owing to their drawbacks, researchers have reported different biomaterials as bioadsorbents for metal ions. Crushed palm fibers and coconut husk can be used as biomaterials to remove chromium(VI) from wastewater. These biomaterials were highly active to biodegrade the metal ions due to their unique elemental composition, such as cellulose, hemicelluloses and lignin. Neem leaf powder can also be used as bioadsorbent to remove metal ions from wastewater due to the presence of active bio-components and amino acids with polar functional groups such as –NH2, –COOH and –OH. These bio-components, along with functional groups, affect the surface properties of neem powder and aid in creating the negative surface charge by increasing the electronegativity [24–26]. Nanoparticles can be prepared from neem powder to enhance the surface chemistry, along with providing more surface area, and thereby increasing the number of exposed active sites. Cellulose and their derivatives have been explored for biosorption to remove metal ions, with remarkable adsorptive capacity. Though the efficacy of cellulose for heavy metal removal has been reported so far, further modifications can be done to improve the ion exchange between cellulose and metals as the adsorption capacity is reduced after each cycle of removal [14, 15, 22, 25, 29].

The reaction proceeds at a faster rate initially due to the high concentration of metals. The percent removal increases by increasing the concentration of metals.

Evidently, this behavior can be directly attributed only due to the fixed and available binding sites. The biosorption increases by increasing the pH of the reaction system. At acidic pH, the metals ions compete for binding sites on the biomaterial. But at alkaline pH, the biosorption decreases as the efficiency of the metal ions binding to the binding sites of biomaterials decreases. The electrostatic interaction between the biomaterials and metal ions is the principal factor to affect the rate of binding between them. The better interaction leads to a higher adsorption of metal ions [12, 13, 53, 55].

Neem powder has showed better efficiency of metals removal due to the presence of high electronegative bio-components. Recently, some novel biomaterials have been reported for metals removal from wastewater by following strategies for both direct and indirect elimination. Direct reduction is a method that involves electron donor groups of biomaterials, and metal atoms are directly reduced. By following the mechanism, Cr(VI) can be reduced to Cr(III) by the functional groups of biomaterials with a lower reduction potential. The indirect reduction mechanism involves the binding of the anionic metal ions to cationic functional groups of biomaterials. The metal ions are thereby reduced due to the adjacent electron donor functional groups and the reduced metal ions are released into the aqueous environment. Sometimes, electronic repulsion is caused due to the cationic functional groups of biomaterials and cationic metals. The reduced metals thereby lead to form complexes through complexation. If the biomaterials have low electron donor functional groups, the metal ions bound to the biomaterials may remain intact and do not reduce. Metals can be removed at a faster rate by increasing the concentration of the biomaterial [8, 38, 51].

As mentioned earlier, the decrease in the pH of the aquatic solution increases the removal rate of metals as the protons (H^+) participate in the anionic metal adsorption and metal reduction reactions. Temperature also affects the removal rate. Usually, the removal rate increases as the temperature increases due to the endothermic redox reactions. However, the acid decreases the active uptake of metal ions due to preferential adsorption of protons, in comparison to metal ions. For example, calcium alginate beads can adsorb metal ions in wastewater in the absence of a biomass. But the rate of metals uptake is increased threefold to fourfold in the presence of biomaterials. It strongly adsorbs metal ions due to the polysilicate layers present on such type of an alginate [27, 28, 46].

Bacteria can also be used as biomaterials to remove metals from wastewater. Different bacteria have been studied and they have recently characterized for their efficiency to remove metal ions. Some examples of such bacteria are *E. coli, P. putida, P. denitrificans* and *B. subtilis*. The ubiquitous methane-oxidizing bacteria can be potentially used for the biodegradation of organic and inorganic contaminants in water. *Methylococcus capsulatus* has been reported to remove chromium metal ions in varying concentrations. The results suggest that the rate of reduction of metals depends on the availability of the reducing components of the substrate (methane) and the active cellular metabolism. The substrate methane is involved in reduction by providing electron donor groups for reduction. Furthermore, the genome sequencing of *M. capsulatus* shows that five genes were involved in the reductase activity against metal ions. *E. coli* has also been reported to reduce metal ions by using NADH (reductant), which reduces metal ions and NADH into NAD-metal complexes [33, 47, 48].

4.3 Conclusion

Metal contamination in wastewater has become one of the most significant environmental concerns around the globe. The supply of potable and safe water has become a major challenge. Different conventional technologies, such as precipitation, coagulation/flocculation, ion exchange, complexation and membrane filtration, have been developed for hazardous, reactive, soft and heavy metals removal. Biosorption is an efficient and cost-effective method to remove metals even at low concentrations. Owing to the successful use of biosorption, an effective biomass with significant biosorption selectivity should be chosen. Biomaterials act as an alternative to the expensive adsorbents for the degradation/removal of metal ion complexes from wastewater. But the standard choice with selectivity and standardization to extract the metal ions needs further research. Future studies are required to fabricate biomaterials for wastewater treatment at a large scale.

References

[1] Huang, G.H. and J. Xia, *Barriers to sustainable water-quality management*. Journal of Environmental Management, 2001. **61**(1): p. 1–23.

[2] Kawawaki, T., et al., *Creation of active water-splitting photocatalysts by controlling cocatalysts using atomically precise metal nanoclusters*. Chemical Communications, 2021. **57**(4): p. 417–440.

[3] Borah, P., M. Kumar and P. Devi. Types of inorganic pollutants: metals/metalloids, acids, and organic forms. In Inorganic pollutants in water (17–31). 2020. Elsevier.

[4] Wasewar, K.L., S. Singh and S.K. Kansal. Process intensification of treatment of inorganic water pollutants. In Inorganic pollutants in water (245–271). 2020. Elsevier.

[5] Wilkie, P.J., G. Hatzimihalis, P. Koutoufides and M.A. Connor. The contribution of domestic sources to levels of key organic and inorganic pollutants in sewage: the case of Melbourne, Australia. Water Science and Technology, 1996. **34**(3–4): p. 63–70.

[6] Bharagava, R.N., G. Saxena and S.I. Mulla. Introduction to industrial wastes containing organic and inorganic pollutants and bioremediation approaches for environmental management. Bioremediation of Industrial Waste for Environmental Safety: Volume I: Industrial Waste and Its Management, 2020: p. 1–18.

[7] Serpone, N., E. Borgarello and E. Pelizzetti. Photoreduction and photodegradation of inorganic pollutants: II. Selective reduction and recovery of Au, Pt, Pd, Rh, Hg, and Pb. Photocatalysis and Environment: Trends and Applications, 1988: p. 527–565.

[8] Serpone, N., E. Borgarello and E. Pelizzetti. Photoreduction and photodegradation of inorganic pollutants: I. Cyanides. Photocatalysis and Environment: Trends and Applications, 1988: p. 499–526.

[9] Wang, K., J. Zhang, L. Lou, S. Yang and Y. Chen. UV or visible light induced photodegradation of AO7 on TiO2 particles: the influence of inorganic anions. Journal of Photochemistry and Photobiology A: Chemistry, 2004. **165**(1–3): p. 201–207.

[10] J Alvarez, M.A., M. Ruidíaz-Martínez, G. Cruz-Quesada, M.V. López-Ramón, J. Rivera-Utrilla, M. Sanchez-Polo and A.J. Mota, Removal of parabens from water by UV-driven advanced oxidation processes. Chemical Engineering Journal, 2020. **379**: p. 122334.

[11] Garg, T., J. Kaur, V. Kumar, K. Tikoo, A. Kaushik and S. Singhal. A novel dualresponsive bio-derived magnetic nanocomposite: An excellent catalyst for efficient photodegradation of contaminants and detection of food additive in wastewater. Chemical Engineering Journal, 2023: p. 144218.

[12] Karthik, R., J. Vinoth Kumar, S.M. Chen, C. Karuppiah, Y.H. Cheng and V. Muthuraj. A study of electrocatalytic and photocatalytic activity of cerium molybdate nanocubes decorated graphene oxide for the sensing and degradation of antibiotic drug chloramphenicol. ACS Applied Materials and Interfaces, 2017. **9**(7): p. 6547–6559.

[13] Ghalkhani, M., N. Zare, F. Karimi, C. Karaman, M. Alizadeh and Y. Vasseghian, Recent advances in Ponceau dyes monitoring as food colorant substances by electrochemical sensors and developed procedures for their removal from real samples. Food and Chemical Toxicology, 2022. **161**: p. 112830.

[14] Thassitou, P.K. and I.S. Arvanitoyannis. Bioremediation: a novel approach to food waste management. Trends in Food Science and Technology, 2001. **12**(5–6): p. 185–196.

[15] Haripriyan, U., K.P. Gopinath, J. Arun and M. Govarthanan. Bioremediation of organic pollutants: A mini review on current and critical strategies for wastewater treatment. Archives of Microbiology, 2022. **204**(5): p. 286.

[16] Mallikarjunaiah, S., M. Pattabhiramaiah and B. Metikurki. Application of nanotechnology in the bioremediation of heavy metals and wastewater management. Nanotechnology for Food, Agriculture, and Environment, 2020: p. 297–321.

[17] Sonune, N.A. and A.M. Garode. Bioremediation potential of bacterial isolates for municipal wastewater treatment. Current world environment, 2015. **10**(2): p. 619–625.

[18] Machałowski, T., M. Wysokowski, I. Petrenko, A. Fursov, M. Rahimi-Nasrabadi, A. Moh'd M, H. Meissner, Y. Joseph, B. Fazilov, H. Ehrlich and T. Jesionowski, Naturally pre-designed biomaterials: Spider molting cuticle as a functional crude oil sorbent. Journal of environmental management, 2020. **261**: p. 110218.

[19] Grabi, H., F. Derridj, W. Lemlikchi and E. Guénin. Studies of the potential of a native natural biosorbent for the elimination of an anionic textile dye Cibacron Blue in aqueous solution. Scientific Reports, 2021. **11**(1): p. 9705.

[20] Lansari, F., M. Edjekouane, O. Khelifi, I. Boukheteche and H. Laksaci. Elimination of methylene blue by low-cost biomaterial prepared from local natural residue. Algerian Journal of Renewable Energy and Sustainable Development, 2020. **2**(1): p. 60–66.

[21] Ghebremichael, K., N. Gebremedhin and G. Amy. Performance of Moringa oliefera as a biosorbent for chromium removal. Water Science and Technology, 2010. **62**(5): p. 1106–1111.

[22] Gupta, B., et al., *Fabrication of calcite based biocomposites for catalytic removal of heavy metals from electroplating industrial effluent*. Environmental Technology & Innovation, 2021. **21**: p. 101278.

[23] Elakkiya, S., et al., *Review on characteristics of biomaterial and nanomaterials based polymeric nanocomposite membranes for seawater treatment application*. Environmental research, 2021: p. 111177.

[24] Chakraborty, V., P. Das, and P.K. Roy, *Synthesis and application of various metal oxide–/biomaterial–coated carbonaceous nanomaterials derived from waste biomass for removal of Cr+ 6 present in solution*. Biomass Conversion and Biorefinery, 2021: p. 1–14.

[25] Tan, X., et al., *High value-added biomaterials recovery from granular sludge based wastewater treatment process*. Resources, Conservation and Recycling, 2021. **169**: p. 105481.

[26] Foroutan, R., et al., *Hydroxyapatite biomaterial production from chicken (femur and beak) and fishbone waste through a chemical less method for Cd2+ removal from shipbuilding wastewater*. Journal of Hazardous Materials, 2021. **413**: p. 125428.

[27] Almomani, F. and R.R. Bohsale, *Bio-sorption of toxic metals from industrial wastewater by algae strains Spirulina platensis and Chlorella vulgaris: Application of isotherm, kinetic models and process optimization*. Science of The Total Environment, 2021. **755**: p. 142654.

[28] Kartik, A., et al., *A critical review on production of biopolymers from algae biomass and their applications*. Bioresource Technology, 2021: p. 124868.

[29] Romero-Soto, I.C., et al., *Galactomannans for Entrapment of Gliomastix murorum Laccase and Their Use in Reactive Blue 2 Decolorization*. Sustainability, 2021. **13**(16): p. 9019.

[30] Akar, T., et al., *Immobilized Mucor plumbeus on sepiolite support: A potential decolorization agent suitable for batch and continuous mode water treatment*. Journal of Cleaner Production, 2021. **294**: p. 126283.

[31] Latif, A., et al., *Immobilization of Trametes Versicolor Laccase on Cu-alginate Beads for Biocatalytic Degradation of Bisphenol A in Water: Optimized Immobilization, Degradation And Toxicity Assessment*. Journal of Environmental Chemical Engineering, 2021: p. 107089.

[32] Chen, X., et al., *Effect of Geobacillus toebii GT-02 addition on composition transformations and microbial community during thermophilic fermentation of bean dregs*. Scientific Reports, 2021. **11**(1): p. 1–14.

[33] Negroiu, M., et al., *Novel Adsorbent Based on Banana Peel Waste for Removal of Heavy Metal Ions from Synthetic Solutions*. Materials, 2021. **14**(14): p. 3946.

[34] Villen-Guzman, M., et al., *Valorization of lemon peel waste as biosorbent for the simultaneous removal of nickel and cadmium from industrial effluents*. Environmental Technology & Innovation, 2021. **21**: p. 101380.

[35] Pavithra, S., et al., *Batch adsorption studies on surface tailored chitosan/orange peel hydrogel composite for the removal of Cr (VI) and Cu (II) ions from synthetic wastewater*. Chemosphere, 2021. **271**: p. 129415.

[36] Shaikhiev, I.G., N.V. Kraysman, and S.V. Sverguzova, *Review of Almond (Prunus Dulcis) Shell Use to Remove Pollutants from Aquatic Environments*. 2021.

[37] Iwuozor, K.O., et al., *A review on the mitigation of heavy metals from aqueous solution using sugarcane bagasse*. Sugar Tech, 2021: p. 1–19.

[38] Assirey, E.A. and L.R. Altamimi, *Chemical analysis of corn cob-based biochar and its role as water decontaminants*. Journal of Taibah University for Science, 2021. **15**(1): p. 111–121.

[39] Al Rmalli, S.W., et al., *A biomaterial based approach for arsenic removal from water*. Journal of environmental monitoring, 2005. **7**(4): p. 279–282.

[40] Elgarahy, A., et al., *A critical review of biosorption of dyes, heavy metals and metalloids from wastewater as an efficient and green process*. Cleaner Engineering and Technology, 2021. **4**: p. 100209.

[41] Elnashar, M.M., *Immobilized molecules using biomaterials and nanobiotechnology*. Journal of Biomaterials and Nanobiotechnology, 2010. **1**(01): p. 61.

[42] Fabrizius-Homan, D. and S. Cooper, *A comparison of the adsorption of three adhesive proteins to biomaterial surfaces*. Journal of Biomaterials Science, Polymer Edition, 1992. **3**(1): p. 27–47.
[43] Landin-Sandoval, V., et al., *Valorization of agri-food industry wastes to prepare adsorbents for heavy metal removal from water*. Journal of Environmental Chemical Engineering, 2020. **8**(5): p. 104067.
[44] Minamisawa, M., et al., *Adsorption behavior of heavy metals on biomaterials*. Journal of agricultural and food chemistry, 2004. **52**(18): p. 5606–5611.
[45] de la Rosa, G., et al., *Recycling poultry feathers for Pb removal from wastewater: kinetic and equilibrium studies*. International Journal of Chemical and Biomolecular Engineering, 2008. **1**(4): p. 185–193.
[46] Orejuela-Escobar, L., et al., *Prospects of Microalgae for Biomaterial Production and Environmental Applications at Biorefineries*. Sustainability, 2021. **13**(6): p. 3063.
[47] Ray, S.S. and A.O.C. Iroegbu, *Nanocellulosics: Benign, sustainable, and ubiquitous biomaterials for water remediation*. ACS omega, 2021. **6**(7): p. 4511–4526.
[48] Feng Su, J., et al., *Multifunctional modified polyvinyl alcohol: A powerful biomaterial for enhancing bioreactor performance in nitrate, Mn (II) and Cd (II) removal*. Water research, 2020. **168**: p. 115152.
[49] Begum, S., et al., *Metal–organic framework-templated biomaterials: Recent progress in synthesis, functionalization, and applications*. Accounts of chemical research, 2019. **52**(6): p. 1598–1610.
[50] Sirajudheen, P., et al., *Applications of chitin and chitosan based biomaterials for the adsorptive removal of textile dyes from water – A comprehensive review*. Carbohydrate Polymers, 2021. **273**: p. 118604.
[51] Aramesh, N., A.R. Bagheri, and M. Bilal, *Chitosan-composite/hybrid biomaterials for adsorptive removal of dyes and underlying interaction mechanisms*. International Journal of Biological Macromolecules, 2021.
[52] Rizal, S., et al., *Cotton Wastes Functionalized Biomaterials from Micro to Nano: A Cleaner Approach for a Sustainable Environmental Application*. Polymers, 2021. **13**(7): p. 1006.
[53] Feng, C., et al., *Extracellular biopolymers recovered as raw biomaterials from waste granular sludge and potential applications: A critical review*. Science of the Total Environment, 2021. **753**: p. 142051.
[54] Fan, H., et al., *Adsorption properties and mechanisms of novel biomaterials from banyan aerial roots via simple modification for ciprofloxacin removal*. Science of The Total Environment, 2020. **708**: p. 134630.
[55] Sun, Y., et al., *Enhanced performance for Hg (II) removal using biomaterial (CMC/gelatin/starch) stabilized FeS nanoparticles: Stabilization effects and removal mechanism*. Chemical Engineering Journal, 2018. **344**: p. 616–624.
[56] Opeolu, B.O., et al., *Utilization of biomaterials as adsorbents for heavy metals removal from aqueous matrices*. Scientific Research and Essays, 2010. **5**(14): p. 1780–1787.

Sadaf Tariq, Awais Ahmad, Rafael Luque, Mabkhoot Alsaiari, Farid A. Harraz

5 Bionanomaterials and their impact on organic pollutants via photodegradation

Abstract: Due to rapid industrialization, organic contaminants are becoming a significant concern with special consideration on hazardous and refractory contaminants that need to be addressed effectively. Biomaterials and nanomaterials can be coupled to develop bionanomaterials for synergistic effect of biosorption and photocatalysis. The coupled bionanomaterials can be effectively used to remove organic pollutants by providing multiple and selective functional groups for efficient absorption on the nanoscale surface of biomaterials. Also, bionanomaterials help in dispersion and recovery of nanomaterials by modifying their band structure. In this chapter, the green route approach is discussed for the synthesis of eco-friendly and cost-effective bionanomaterials. The fabrication of biomaterials/nanomaterials is also discussed. The application of fabricated bionanomaterials to remove organic pollutants via photodegradation is also addressed. Although bionanomaterials are efficient to remove pollutants, there are still a few bottlenecks for their limited efficiency. Therefore, we also present an up-to-date research progress on bionanomaterials to improve their pollutant removal efficiency.

Keywords: Bionanomaterial, biomaterial, pollutants, biodegradation, photocatalysis

5.1 Introduction

Over several decades, the environmental pollution has become a serious concern due to industrialization across the globe. Different pollutants such as metals, organic pollutants and pathogens from industrial wastes are released into to either terrestrial or aquatic environment. These pollutants are nonbiodegradable and are a constant threat to living organisms and their environment even in trace amounts [1–3]. The organic pollutants are removed from wastes of several industries such as fertilizers, pharmaceuticals, dyes, mining and pesticides. They prefer to concentrate in the food chain and are poisonous and nonbiodegradable, may be carcinogenic thus imposing terrific health risks. These organic pollutants are commonly characterized by their wide dispersion in aquatic and terrestrial ecosystems due to their easy bioaccumulation. They can cause chronic toxicity, cancer, endocrine disruption and drug resistance. Owing to their tremendous health concerns, it is necessary to develop some effective strategies for complete and effective removal of organic pollutants from the environment [4–6].

https://doi.org/10.1515/9783110768749-005

Various conventional materials such as adsorbents are in use to remove pollutants from environment. Mostly, the activated carbon is used; however, its cost-ineffectiveness has urged researchers to find alternatives of same activity. Biomaterials have emerged as an alternative to conventional adsorbents with their environment benignity and cost-effective properties. Some biomaterials are used as photocatalysts to trigger redox reaction in organic pollutants as shown in Figure 5.1 [7–10]. The process of photocatalysis utilizes solar light energy and generates charge carriers of high energy for induction of redox reactions. In this regard, the fabrication of biomaterials and nanomaterials can be an effective and promising strategy due to their combined and synergistic activities such as nontoxicity, high durability, enhanced activity, low cost and improved stability. Various semiconductors have been fabricated with biomaterials at nanoscale to utilize them for cleaning of environment. The photodegradation property of bionanomaterials has been successfully employed to remove various pollutants. The organic pollutants are reduced due to the high energetic electrons generated by photons of solar light and can be either removed or recovered [11–14].

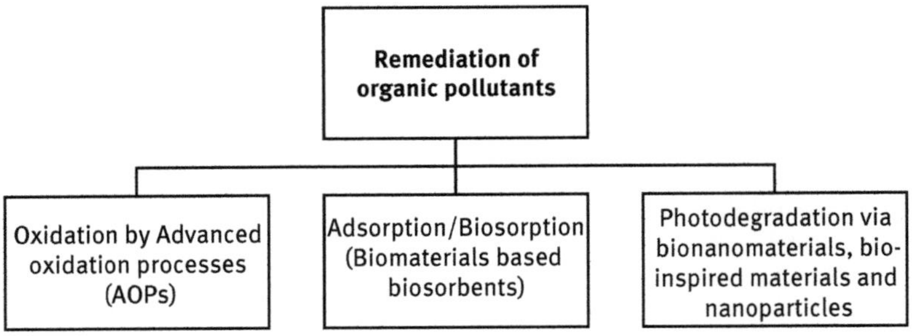

Figure 5.1: Commonly used strategies for the remediation of organic pollutants.

The photon-mediated degradation of pollutants by bionanomaterials has been used as a promising strategy to turn waste into treasure, and the recovered material can be further used in various fields. For example, degradation of organic contaminants into simpler inorganic compounds is possible due to oxidation of parent molecules by reactive oxygen species. Also, the treatment of contaminants by bare and pure biomaterials is not sufficient; therefore, it is necessary to fabricate bionanomaterials for light-driven redox reactions on their surfaces for enhanced photodegradation [15–18].

Another aspect is that green chemistry approaches can be used for the synthesis of bionanomaterials which helps their dispersion and recovery from the reaction system after bioremediation in a portable platform. Only the appropriately and well-designed bionanomaterial is so efficient to retain organic pollutants to its surface and increase the efficiency of a photocatalyst. So, the proper immobilization of bionanomaterials by using living or nonliving biomaterials will not only enhance the efficiency of bionanomaterials to remove pollutants but also increase their photodegradation. Thus, by developing and modifying biomaterial/nanomaterial composites as photocatalysts can be selectively and efficiently used to remove trace organic pollutants, thereby converting and obtaining the organic–inorganic composites to exhibit enhanced photodegradation of pollutants [19–21]. This chapter aims to summarize the impact of bionanomaterials on photodegradation of organic pollutants. Furthermore, the recent advances in the fabrication of bionanomaterials as photocatalysts and their applications in environment cleaning are also provided. The subsequent sections of the chapter also brief an insight to improve the fabrication and development of bionanomaterials by green chemistry approach for photodegradation.

5.2 Bionanomaterials for photodegradation of organic pollutants

Different nanomaterials, graphene oxides, carbon nanotubes, covalent organic frameworks and MXenes can be used to degrade pollutants from the environment with excellent properties such as high biosorption, selectivity and stability as shown in Table 5.1. Owing to the biocompatibility and complementarity, any biomaterial should be able to recognize the organic pollutant molecules effectively. Only the biomimetic materials can recognize pollutants selectively to adsorb them on their surfaces. For example, the TiO_2 loaded on eggshell has been reported to remove pollutants with very high efficiency due to the synergy between egg shell (biomaterials) and TiO_2 photocatalyst [22, 23]. Other biomaterials such as fish scales can also be used as biosorbents owing to their binding capacity. The fish scales are rich in collagen protein with alluring functional groups, including phosphate, carboxyl, thiol, amine and amide for biosorption of pollutants. Various carbon sources have also been reported for their enhanced activity in photodegradation of pollutants [24, 25].

The photodegradation of persistent and synthetic organic pollutants under the radiations of visible light is one of the most promising ways to remove pollutants at even trace concentrations efficiently. TiO_2 and ZnO coupled with different biomaterials have been extensively reported by researchers for photodegradation of organic pollutants. MXene is an efficient photocatalyst with facilitated separation, generation and recombination of electron–hole pairs. The modified Ti_3C_2–MXene/Bi_xFe_y nanocomposites have been synthesized to degrade Congo red and acetophenone in solar light. The Congo red was degraded within 42 min and acetophenone within 150 min. Owing to the metal-free, low-cost and ubiquitous nature of C_3N_4 has been significantly used for photodegradation of various organic and inorganic pollutants

Table 5.1: Summary of different nanomaterials used for removal of pollutants.

Nanomaterials	Merits	Demerits
Graphene oxides	Improved biosorption, surface functionalization, functional groups and large surface area	Difficulty in separation from reaction system, poor selectivity and cannot be used on a large scale
Carbon nanotubes	Large surface area, highly stable, low pH working ability and easy synthesis	Relatively high cost, low selectivity, low biosorption ability and cannot be used on large-scale synthesis
Metal organic frameworks	Large-scale synthesis, functional group, large specific surface area and high porosity	High cost and low hydrophilicity
Covalent organic frameworks	Easy biosorption, highly thermally and chemically stable, functional groups and surface functionalization	Costly synthesis, difficult separation and control of structure
MXenes	Accessible and excessive biosorption sites, and controlled layered structure	High cost and poor selectivity
Bionanomaterials	Photodegradation of organic pollutants even at low concentration, highly selective, highly stable, cost-effective and environment benignity	Low efficiency after continuous use

as an efficient photocatalyst. The $AlVO_4/g\text{-}C_3N_4$ nanohybrids can be used for photodegradation of methylene blue from wastewater due to the significant generation of electron–hole pair in response to visible and ultraviolet light. Additionally, the role of $TiO_2/g\text{-}C_3N_4$ has also been reported for the photodegradation of acetaminophen under visible light irradiation. The as-obtained nanocomposites exhibit stronger and selective oxidation of pollutants, thereby degrading and removing pollutants from the environment [26, 27].

The Ag nanoparticles decorated on phosphorus-doped C_3N_4 surface have been reported for photodegradation of sulfamethoxazole with high stability and durability for at least six cycles without any loss in photocatalytic activity. These nanocomposites have also been used for removal of dyes with ~98% efficiency and up to 33% efficiency for removal of metals. The photodegradation of pollutants can be ascertained by radical scavengers by contributing active species in the reaction system. Hence, the heteroatom doping can improve the absorption of light, enhance the separation of electron–hole pairs, introduce multiple electron-reactive centers to trap the substrates and increase the transfer of charges [28–30].

Therefore, the recycling and maintaining the stability of as-synthesized photocatalyst can be used as a strategy to improve their applications in the photodegradation of organic pollutants. The photodegradation of pollutants is appropriate for the treatment of wastewater at trace concentrations. At high concentrations, the organic pollutants may need longer duration for their complete removal and degradation [10, 31, 32].

5.2.1 Biochar-based bionanomaterials

Biochar-based bionanomaterials act as a surrogate for conventional materials. Under pyrolysis and carbonization, the dead biomass or carbon-rich biomaterial such as wood, agro-by-products, sugarcane bagasse and livestock manure can be converted into biochar. The biochar can be used as a precursor to improve the adsorptive properties [33]. The availability of excessive functional groups on the surface make them advantageous over conventional activated substances as these groups exhibit alluring catalytic properties and serve as a platform for biosynthesis of biochar-based bionanomaterials. The structural and chemical properties define the extent of interaction between the biochar and pollutants. Different mechanisms can be followed by biochars for removal of pollutants such as pore-filling, diffusion, partitioning, surface hydrophobic and hydrophilic interactions, aromatic interactions, cationic interactions, electrostatic interactions and hydrogen bonding [31]. The uptake of organic pollutants by biochars is not only the function of surface but also the bulk properties of carbonized and non-carbonized fractions. The photocatalytic oxidation of pollutants can be enhanced by developing surface-modified biochars using corn cob as a precursor via sol–gel method. Surprisingly, the as-obtained biochar exhibited higher removal efficiency due to hydrophobic interaction between the biochars and pollutants than other pure counterparts. The photocatalytic oxidation of organic pollutants by using biochars depends upon the loading and irradiation time with negligible toxicity which confirms the stability of biochar-based bionanomaterials and their eco-friendly heterogeneous photocatalysis [32, 34, 35].

5.2.2 Lignin-based bionanomaterials

The lignin can be employed as a precursor and biomaterial for the synthesis of mesoporous nanomaterials through hydrolysis precipitation approach. Owing to the electronegativity and unique structural skeleton, the as-obtained nanocomposites possess a BET surface area and remarkable porosity. Due to the enhanced interactions between functional groups of biomaterial and the surface functional groups of the nanoparticles, the calcined composites afford better crystallinity and mesoporous structure along with remarkable particle growth. However, the doped metal atoms are low electronegative and therefore easily react with a nucleophile [36, 37]. The polycyclic mesh of lignin has electronegative -OH groups with stronger affinity for cationic-doped metal ions. Other functional groups of lignin, such as phenolic, alcoholic and carboxyl groups, can also react with biopolymers to form nanocomposites.

The as-obtained bionanomaterials exhibit significant efficiency due to their proper fabrication and uniform distribution of nanopores. Due to the nonporous nature of polysaccharides, they are limited in their diffusion and surface property-related applications. To obtain mesoporous polysaccharides, certain presynthesis and postsynthesis modifications such as preheating can be done prior to the formation of gel. The surface tension can be lowered by exchanging water with ethanol. The postsynthetic modification such as carbonization of obtained material for the synthesis of final mesoporous carbons can be done. The controlled carbonized polysaccharides show unique chemical and structural properties which depend upon their degree of carbonization. Another important property is their outstanding mesoporous nature with remarkable porosity [29, 38–40].

5.2.3 Other bionanomaterials for photodegradation

Natural gum, as a polysaccharide, has antimicrobial properties with high binding efficiency. Owing to its unique properties and classification, natural gum-based nanocomposites can be effectively used to remove organic pollutants. The iron-containing guar gum bionanomaterials have been fabricated and used for photodegradation reaction of organic pollutants under solar light illumination [41]. The metal oxide nanoparticles have also been fabricated with natural gum to enhance their efficiency, as a bionanomaterial such as Al_2O_3 powder supported on guar gum has been reported for 90% photodegradation of malachite green when compared to pure Al_2O_3. The xanthan gum is another class of natural gum, and it has been studied for its photodegradation activity against organic pollutants. TiO_2-based bionanocomposites have been synthesized and used for photodegradation under visible light as shown in Table 5.2 [42, 43]. The large exposed surface area and photon-harvesting property of xanthan gum–TiO_2 increased the photodegradation efficiency of composites to more than 74% as compared to pure counterparts. Other modifications such as addition of hydrogen peroxide in the reaction system enhanced the removal efficiency of composites to 89%. The gum arabic and karaya gum can also be used to form natural gum nanocomposites such as TiO_2/ gum arabic and TiO_2/karaya gum have been reported for better photodegradation of organic pollutants under visible light irradiation due to the reduced bandgap of TiO_2. Due to their hierarchical biodegradation and biocompatible nature, these nanocomposites can be used as photocatalysts and bionanomaterials to degrade and remove organic pollutants of various nature from different ecosystems [42, 44, 45].

Collagen is a protein and can be used for biodegradation of various pollutants. The modified and cationic collagen protein can be immobilized on Fe^{3+}-doped metal oxide nanoparticles as a novel photocatalyst. Particularly, high weight percentile collagen can be loaded on iron-doped TiO_2 nanocomposites for maximal photodegradation [55, 56]. The iron-doped TiO_2 cationic collagen has

Table 5.2: Summary of bionanomaterials used for photodegradation of organic pollutants.

Bionanomaterials	Organic pollutants	Intensity of light	Photodegradation efficiency	Ref.
TiO_2/chitosan/Ag	4-Nitrophenol	UV	100%	[46]
TiO_2/chitosan/glass	Methyl orange	White light	87%	[47]
Core–shell/chitosan/TiO_2	MO	UV	55%	[48]
Cs–TiO_2/glass	MO	UV	White light, UV	[49]
Wool fabrics–TiO_2	Methylene blue	UV	6%	[50]
TiO_2/chitosan matrix	Rhodamine B	White light	94%	[51]
Aegle marmelos leaf–Ag–Au–TiO_2	SDBS	UV	81%	[52]
TiO_2/Fe–N–cotton knit	Methylene blue	UV	87%	[53]
Chitosan–TiO_2	Rhodamine B	UV	94%	[53]
Cellulose/titania-based films	Methylene blue	Solar light	80%	[50]
Chitosan–MoO_3–TiO_2	Methyl orange	Solar light	70%	[54]

been reported for high degradation efficiency of ~96% under visible light irradiation due to $-N(CH_3)^{3+}$ functional group, surface immobilization, iron doping and nanoparticle surface. Owing to these properties, the nanocomposites promote the rapid migration of anionic organic pollutants on the surface of iron-doped TiO_2 nanomaterial through strong electrostatic interactions and thereby establish an adsorption equilibrium. The adsorbed organic molecules degrade at the surface rapidly. Additionally, the sites of active adsorption act as trap centers for induced electrons generated by irradiated photons of light. The collagen monohybrids can also be supported on different silver salts to prepare nanocomposites as bionanomaterials for highly efficient photodegradation of organic pollutants [57–59].

The silver acetate, silver chloride, silver phosphate and silver molybdate can be immobilized on collagen nanohybrids through in situ synthesis. Their photodegradation efficiency has been reported against methyl orange under ultraviolet and visible light illumination. The surface plasmon resonance effect of nanoparticles can be related to the efficiency to absorb light and generation of electron–hole pairs. The induced electrons are shifted to conduction band and react with pollutants to form superoxide radicals. Also, the generated holes of valence band are migrated to its surface and react with nanoparticles, thereby decomposing the organic pollutants at the surface of bionanomaterials [30, 60, 61].

5.3 Fabrication of biomaterials/nanomaterials

The biomaterials can be fabricated with nanomaterials for bioremediation of environmental pollutants with distinctive qualities like biodegradable, biocompatible, nontoxic and low-cost fabricated materials. These materials act as bioadsorbents for organic pollutants in which biomaterials act as a support to disperse nanoparticles. By using well-developed procedures, the multifunctional bionanomaterials can be obtained. The biomaterials/nanomaterials are fabricated to obtain the synergistic effect. The inherited multifunctional biomaterials provide support for the dispersion of nanoparticles, enhance the overall stability of nanoparticles and make it reusable for multiple cycles of activity [1, 31, 40, 43]. The adsorption capacity of biomaterials along with their photocatalytic property of nanoparticles can be integrated to form bionanomaterials, which can remove various organic pollutants even in trace amounts. Under specific environmental and reaction conditions, the band structure can also be tailored to improve the absorption of solar light and separation of photogenerated charge carriers. Furthermore, the metals, nonmetals, heteroatoms and metal oxides can be doped to enhance the photodegradation activity of bionanomaterials under radiations of visible light [23]. Along with the synergistic effect induced by these materials, other functional groups can also be responsible for effective absorption of molecules of pollutants by them. The ionic/molecular imprinting of these materials enhances their selectivity toward any specific organic pollutant. It makes easier for the biomaterial to separate and recover pollutants by the coupled effect of magnetic Fe_3O_4 [10, 31–33].

The synergy between biomaterials and nanomaterial specifically depends upon the obtained material and the extent of formed interaction between them. Up to now, various researches reported the facile synthesis of bionanomaterials via simple mixing of nanomaterials with the biomaterials. Additionally, some postsynthetic modifications can be done to bionanomaterials to obtain supported or free-standing nanostructures. The restricted use of nanomaterials and controlled synthesis can be employed to obtain ordered nanofilms of photodegradation activity. The development of biomaterials/TiO_2 nanocomposites as photocatalysts with core–shell morphology has been reported recently with maximum exposure of active sites onto its surface to enhance the biosorption of pollutants along with photodegradation of organic pollutants under exposure to visible light [9, 44, 45, 61].

5.3.1 Mechanism of action

To understand the basic mechanism of photodegradation via bionanomaterials, different spectroscopic techniques, theoretical calculations and characterization techniques can be studied. Density functional theory (DFT) is a very useful tool to provide data about the exposed and accessible active sites, the nature of substrates, total energy and charge density. The interaction between bionanomaterials and pollutants can be understood by analysis of the electronic properties. The reaction intermediates, transition states, reaction conditions, oxidation states and bond distances of the photocatalytic degradation can be stimulated by further characterization studies. The DFT calculation of photodegradation can be used to predict the state of reaction intermediates, their transition states, oxidation states and bond energies as an evidence of the possible pathway degradation. The X-ray photoelectron spectroscopy analysis can be done to obtain information about the chemical composition, oxidation state, concentration and bonding relationship of targeted ions [50, 62–64].

5.3.2 Organic polymers as nanomaterials

Organic pollutants are recalcitrant and highly persistent in the environment to cause serious health threats. These organic pollutants released from pharmaceuticals, domestic products, synthetic chemicals, pesticides and organic dyes can be remediated in wastewater. The synthetic and organic dyes are highly carcinogenic and can be divided into cationic and anionic dyes. Certain bionanomaterials with high surface-to -volume ratio can be extensively used to degrade these pollutants from wastewater. For example, silver bromide-based nanomaterials supported on hexadecyltrimethyl ammonium bromide has showed efficient activity to remove organic dyes such as Congo red, orange, indigo carmine, methylene blue and methyl orange in wastewater under visible light irradiations. Photodegradation acts as a useful remedy for bioremediation without any side effect to the environment [29, 38–40].

The organic polymers can be used as nanomaterials in the form of pure organic-based nanoparticles and inorganic–organic polymers such as cellulose and chitosan. Owing to the presence of $-NH_2$ and $-OH$ functional groups, the biomaterial-based nanomaterials have gained much attention for bioremediation of wastewater. The organic and inorganic pollutants can be chelated and removed from the environment. The incorporation of inorganic nanomaterials with biopolymers affords an alternative method by fabricating novel nanocomposites with improved catalytic efficiency as compared to other pure counterparts. The inorganic nanomaterials such as carbon-based supports, carbon nanotubes and graphene act as a support to develop composites of biopolymers by using biomaterials such as chitosan, lignin and cellulose. The hybrids of biomaterials and nanocomposites can be used as a support to stabilize transition metals on nanoparticles [50, 54].

5.4 Green synthesis of bionanomaterials

Toward the safe synthesis of bionanomaterials, researchers are focusing on the significance of green synthesis methods over other chemical-based methods. Conventional synthesis methods involve harsh reaction conditions, complex synthesis steps, high-tech apparatus and hazardous synthetic chemicals, whereas the green synthesis methods are highly favorable, facile, convenient, cost-effective, energy-effective and use renewable resources. These steps can be scaled up and modified with least synthesis of toxic by-products [5, 28, 65].

Over the last decade, various bionanomaterials have been fabricated via green synthesis methods for photodegradation of organic pollutants. These bionanomaterials can be obtained from a variety of living and nonliving biological sources such as plants, microorganisms, agro-by-products, animals, biopolymers, zeolites and dead biomass. Photodegradation ability of bionanomaterials to oxidize organic pollutants under solar light irradiation has also been investigated. The plant-based synthesis of bionanomaterials has been the most frequently used approach to synthesize ZnO-based bionanomaterials. The plant extracts act as excellent reducing agents and stabilizers to remove pollutants due to the abundant phytochemicals in plants. Also, their abundance in nature and low cost make them preferably used as a precursor for development of bionanomaterials which makes them possible and sustainable for large-scale synthesis of bionanomaterials. The bacteria-based synthesis of bionanomaterials using *E. coli* has also been reported for the synthesis of ZnO-based photocatalysts [65–67].

Upon interaction with environment, the bacteria release certain proteins and biologically active molecules not only to protect themselves from environmental conditions but also for the degradation of pollutants. The stress-induced generation and synthesis of bioactive compounds induce the degradation of organic pollutants under visible light irradiation. Other nanomaterial-based photocatalysts have been successfully reported to degrade pollutants such as algae. Other pure biocompounds, including bioflavonoids, rutin, ascorbic acid, vitamins and glycine, act as stabilizers and capping agents to improve the catalytic efficiency of certain bioinspired photocatalysts. In the meantime, the biological compounds act as reducing agents to facilitate the synthesis of bionanomaterials using green synthesis approach [5, 28, 65, 68].

The physicochemical properties of bionanomaterials can be tailored through structural, physical, chemical and electronic modifications to improve photodegradation efficiency. These tailored properties directly affect the characteristics of as-obtained bionanomaterials such as crystallinity, bandgap and concentration. The defects in crystallinity of the obtained materials and vacancies can be attributed as an important which play effective roles in determining their efficiency as a photocatalyst. During photodegradation, these defects and vacancies trap the electrons generated by photons of light to facilitate the separation of charge carriers. All of these processes in combination with other properties enhance the overall activity of photocatalyst. The nanoparticles with high crystallinity and surface area have been reported for more frequent defects as compared to their pure counterparts. These properties can also be tuned and tailored by green synthesis method [9, 44, 45, 61].

To generate high-performance photocatalysts driven by solar light, the metal-based nanomaterials can be doped and coupled with other biomaterials by green synthesis approach as summarized in Table 5.3. Certain metals, metal oxides, non-metals, rare earth metals, alkaline metals and semiconductors have been reported to obtain coupled metal bionanoparticles. These materials are doped to expand the absorption spectrum and reduce the bandgap by increasing the overall efficiency of photocatalyst to absorb more light. Also, the heteroatom doping can be done by green methods to form new electron states in every bandgap to trap the charge carriers generated by photons, which results in retarded recombination of electron–hole pairs. So, by enhancing the quantum efficiency of the photocatalyst, the photodegradation of pollutants can be improved [3, 6, 68].

Table 5.3: Summary of biomaterials synthesized by green synthesis approach.

Biomaterial	Synthesis media	Photodegradation of organic pollutant	Ref.
Aqueous extract of oak fruit hull	Co-reduction method	Basic violet 3 dyes under solar light	[69]
Calotropis gigantea leaves extract	Solution combustion method	Rhodamine B dye under ultraviolet light	[70]
Caralluma fimbriata extract	Simple single-step synthesis	Indigo carmine dye degradation under solar light	[71]
E. coli nitrate reductase enzyme	Simple enzymatic method	Reactive black B dye under solar light	[72]
Leaf extract of *Ocimum basilicum L.*	Solid-state reaction route and modified sol–gel method	Textile effluents under solar light	[73]
Aloe vera gel as fuel	Phytomediated combustion method	Malachite green degradation under light	[74]
Azadirachta indica gum	Solution combustion route	MB photodegradation	[75]

5.5 Conclusion

Herein, the chapter summarizes recent advances in the field of bionanomaterials for photodegradation of organic pollutants by using solar light. A better understanding of various types of mechanisms of bionanomaterials based on their biosynthesis production is provided. The green synthesis method for preparation of bionanomaterials exhibits enhanced photodegradation of pollutants due to their narrow bandgap, high adsorption and improved porosity. Additionally, the green synthesized bionanomaterials possess antimicrobial property in the environment treatment. The bionanomaterials are useful for effective removal of organic pollutants via photodegradation due to high functionality, enhanced reactivity, large surface-area-to-volume ratio, antimicrobial activity in wastewater treatment, biocompatible nature, easy processing, stability of nanoparticles and the availability of electron trapping centers for generation of electron–hole pairs. Owing to all these advantages and properties of bionanomaterials, there is still lack of common and large-scale usage of bionanomaterials due to certain drawbacks such as practical uneasiness in their recovery from the reaction system, limited penetration and absorption of light penetration due to surface-adsorbed organic pollutants, opacity of the obtained slurry during photodegradation and unavailability of huge quantity of biomass. Nonetheless, more in-depth study is required to understand the efficient development and fabrication of high-performance bionanomaterials by using green approaches as an effort to clean environment.

References

[1] Khan, M.A. and A.M. Ghouri. Environmental pollution: its effects on life and its remedies. Researcher World: Journal of Arts, Science & Commerce, 2011. **2**(2): p. 276–285.

[2] Alam, G.J. Environmental pollution of Bangladesh—it's effect and control. Pulp and Paper, 2009. **51**(13.17).

[3] Fox, J., J. McKnight, Y. Sun, D. Maung and R. Crawfis, Using a serious game to communicate risk and minimize psychological distance regarding environmental pollution. Telematics and Informatics, 2020. **46**: p. 101320.

[4] Kaur, K., R. Badru, P.P. Singh and S. Kaushal. Photodegradation of organic pollutants using heterojunctions: A review. Journal of Environmental Chemical Engineering, 2020. **8**(2): p. 103666.

[5] Aravind, M., et al., Critical green routing synthesis of silver NPs using jasmine flower extract for biological activities and photocatalytical degradation of methylene blue. Journal of Environmental Chemical Engineering, 2021. **9**(1): p. 104877.

[6] Wang, C.C., J.R. Li, X.L. Lv, Y.Q. Zhang and G. Guo. Photocatalytic organic pollutants degradation in metal–organic frameworks. Energy & Environmental Science, 2014. **7**(9): p. 2831–2867.

[7] Kashif, M., et al., *Effect of potassium permanganate on morphological, structural and electro-optical properties of graphene oxide thin films*. Arabian Journal of Chemistry, 2021. **14**(2): p. 102953.

[8] Waseem, M., A. Ahmad, M. Sagir, U. Younas, Z. Saeed, M. Pervaiz, F. Ali, A.M. Aljuwayid, M.A. Habila and R.R. Karri. Hydrothermal synthesis of V2O5/TiO2 decorated graphitic carbon nitride nanocomposite for biomolecule oxidation inhibition and dye degradation applications. Environmental Research, 2023: p. 116440

[9] Khalid, A., P. Ahmad, M. Uddin Khandaker, Y. Modafer, H.A. Almukhlifi, A.S. Bazaid, A. Aldarhami, A.M. Alanazi, O.A. Jefri, M.M. Uddin and H. Qanash. Biologically Reduced Zinc Oxide Nanosheets Using Phyllanthus emblica Plant Extract for Antibacterial and Dye Degradation Studies. Journal of Chemistry, 2023.

[10] Ali, F., S. Mehmood, A. Ashraf, A. Saleem, U. Younas, A. Ahmad, M.P. Bhatti, G.E. Eldesoky, A. M. Aljuwayid, M.A. Habila and A. Bokhari. Ag–Cu Embedded SDS Nanoparticles for Efficient Removal of Toxic Organic Dyes from Water Medium. Industrial and Engineering Chemistry Research, 2023. **62**(11): p. 4765–4777.

[11] Cai, Y., F. Luo, Y. Guo, F. Guo, W. Shi and S. Yang. Near-infrared light driven ZnIn2S4-based photocatalysts for environmental and energy applications: Progress and perspectives. Molecules, 2023. **28**(5): p. 2142.

[12] Chen, S., J. Wei, X. Ren, K. Song, J. Sun, F. Bai and S. Tian. Recent Progress in Porphyrin/g-C3N4 Composite Photocatalysts for Solar Energy Utilization and Conversion. Molecules, 2023. **28**(11): p. 4283.

[13] Fan, X., X. Liu and Y. Wang, Low-cost and resource-efficient monolithic photocatalyst with enhanced solar light utilization for the photocatalytic treatment of organic wastewater. Chemosphere, 2023. **312**: p. 137052.

[14] Ahmad, A., et al., A novel study on synthesis of egg shell based activated carbon for degradation of methylene blue via photocatalysis. Arabian Journal of Chemistry, 2020. **13**(12): p. 8717–8722.

[15] Gangwar, J., J.K. Sebastian and P. Chandran. Advanced Bionanomaterials for Environmental Remediation. Sustainable Nanomaterials for Biosystems Engineering: Trends in Renewable Energy, Environment, and Agriculture, 2023: p. 193

[16] Afshan, N., A. Bari, A. Khan and Y. Shah. Advanced Bionanomaterials Used to Recover Pollutants From Wastewater. Sustainable Nanomaterials for Biosystems Engineering: Trends in Renewable Energy, Environment, and Agriculture, 2023: p. 271

[17] Afshan, N. and A. Bari. Advanced Bionanomaterials for Environmental Clean Recovery Processes. Sustainable Nanomaterials for Biosystems Engineering: Trends in Renewable Energy, Environment, and Agriculture, 2023: p. 213

[18] Singh, K.R., G. Nagpure, J. Singh and R.P. Singh. Introduction: utility of nanobiotechnology for bioremediation. In Nanobiotechnology for Bioremediation (1–21). 2023. Elsevier.

[19] Zarrabi, A.. Biological applications of green bionanomaterials: diagnosis applications. In Synthesis of Bionanomaterials for Biomedical Applications (443–467). 2023. Elsevier.

[20] Mukhtar, M., M. Zeeshan, M. Saba, A. Saghir and R. Ayub. Functionalization of green synthesized bionanomaterials. In Synthesis of Bionanomaterials for Biomedical Applications (187–207). 2023. Elsevier.

[21] Afshan, N. and A. Bari. Advanced Bionanomaterials for Environmental Clean Recovery Processes. Sustainable Nanomaterials for Biosystems Engineering: Trends in Renewable Energy, Environment, and Agriculture, 2023: p. 213

[22] Kasinathan, K., et al., *Photodegradation of organic pollutants RhB dye using UV simulated sunlight on ceria based TiO_2 nanomaterials for antibacterial applications*. Scientific reports, 2016. **6**(1): p. 1–12.

[23] Lu, F. and D. Astruc, *Nanocatalysts and other nanomaterials for water remediation from organic pollutants*. Coordination Chemistry Reviews, 2020. **408**: p. 213180.

[24] Alharbi, N.S., et al., *Efficient elimination of environmental pollutants through sorption-reduction and photocatalytic degradation using nanomaterials*. Frontiers of Chemical Science and Engineering, 2020: p. 1–12.

[25] Dasineh Khiavi, N., et al., *Visible light driven heterojunction photocatalyst of CuO–Cu2O thin films for photocatalytic degradation of organic pollutants*. Nanomaterials, 2019. **9**(7): p. 1011.

[26] Feng, X., et al., *Review MXenes as a new type of nanomaterial for environmental applications in the photocatalytic degradation of water pollutants*. Ceramics International, 2021. **47**(6): p. 7321–7343.
[27] Tunesi, M.M., et al., *Application of MXenes in environmental remediation technologies*. Nano Convergence, 2021. **8**(1): p. 1–19.
[28] Ganesh, M., et al., *Hydnocarpus alpina Wt extract mediated green synthesis of ZnO nanoparticle and screening of its anti-microbial, free radical scavenging, and photocatalytic activity*. Biocatalysis and Agricultural Biotechnology, 2019. **19**: p. 101129.
[29] Gómez-Avilés, A., et al., *C-modified TiO2 using lignin as carbon precursor for the solar photocatalytic degradation of acetaminophen*. Chemical Engineering Journal, 2019. **358**: p. 1574–1582.
[30] Heng, J.Z.X., et al., *Solar-powered photodegradation of pollutant dyes using silver-embedded porous TiO2 nanofibers*. Nanomaterials, 2021. **11**(4): p. 856.
[31] Wang, T., et al., *Adsorption and visible-light photocatalytic degradation of organic pollutants by functionalized biochar: Role of iodine doping and reactive species*. Environmental Research, 2021. **197**: p. 111026.
[32] Wan, D., et al., *Photogeneration of reactive species from biochar-derived dissolved black carbon for the degradation of amine and phenolic pollutants*. Environmental Science & Technology, 2021. **55**(13): p. 8866–8876.
[33] Ahmaruzzaman, M., *Biochar based nanocomposites for photocatalytic degradation of emerging organic pollutants from water and wastewater*. Materials Research Bulletin, 2021. **140**: p. 111262.
[34] Cai, X., et al., *Titanium dioxide-coated biochar composites as adsorptive and photocatalytic degradation materials for the removal of aqueous organic pollutants*. Journal of Chemical Technology & Biotechnology, 2018. **93**(3): p. 783–791.
[35] Hu, B., et al., *Efficient elimination of organic and inorganic pollutants by biochar and biochar-based materials*. Biochar, 2020. **2**(1): p. 47–64.
[36] Chen, X.-Y., Q.-J. Pan, and Y.-R. Guo, *One-pot synthesis of the g-C3N4/S-doped ZnO composite with assistance of sodium lignosulfonate and its enhanced photodegradation on organic pollutants*. Materials Research Bulletin, 2018. **107**: p. 164–170.
[37] Chen, X., et al., *Synthesis of visible light responsive iodine-doped mesoporous TiO2 by using biological renewable lignin as template for degradation of toxic organic pollutants*. Applied Catalysis B: Environmental, 2019. **252**: p. 152–163.
[38] Chen, J., et al., *Thermally-assisted photodegradation of lignin by TiO_2/H_2O_2 under visible/near-infrared light irradiation*. Science China Materials, 2018. **61**(3): p. 382–390.
[39] Khan, A., J.C. Colmenares, and R. Gläser, *Lignin-based composite materials for photocatalysis and photovoltaics*. Lignin Chemistry, 2020: p. 1–31.
[40] Arutanti, O., et al., *Advanced Degradation of Lignin from Palm Oil Mill Effluent (POME) by a Combination of Photocatalytic-Fenton Treatment and TiO_2 Nanoparticle as the Catalyst*. Water, Air, & Soil Pollution, 2020. **231**: p. 1–10.
[41] Pathania, D., et al., *Novel guar gum/Al2O3 nanocomposite as an effective photocatalyst for the degradation of malachite green dye*. International journal of biological macromolecules, 2016. **87**: p. 366–374.
[42] Rani, M. and U. Shanker, *Remediation of organic pollutants by potential functionalized nanomaterials*, in *Handbook of Functionalized Nanomaterials for Industrial Applications*. 2020, Elsevier. p. 327–398.
[43] Zhao, Q., X. Zhao, and J. Cao, *Advanced nanomaterials for degrading persistent organic pollutants*, in *Advanced Nanomaterials for Pollutant Sensing and Environmental Catalysis*. 2020, Elsevier. p. 249–305.

[44] Santoso, S.P., et al., *TiO2/guar gum hydrogel composite for adsorption and photodegradation of methylene blue*. International Journal of Biological Macromolecules, 2021. **193**: p. 721–733.
[45] Rani, M., K. Keshu, and U. Shanker, *Sunlight induced photocatalytic degradation of organic pollutants by biosynthesized heterometallic oxides nanoparticles*. Environmental Science and Pollution Research, 2021.
[46] Bergamonti, L., et al., *3D printed chitosan scaffolds: A new TiO2 support for the photocatalytic degradation of amoxicillin in water*. Water Research, 2019. **163**: p. 114841.
[47] Amir, M.N.I., N. Muhd Julkapli, and S.B.A. Hamid, *Effective adsorption and photodegradation of methyl orange by TiO2-chitosan supported glass plate photocatalysis*. Materials Technology, 2017. **32**(4): p. 256–264.
[48] Shi, X., et al., *TiO2-doped chitosan microspheres supported on cellulose acetate fibers for adsorption and photocatalytic degradation of methyl orange*. Polymers, 2019. **11**(8): p. 1293.
[49] Habiba, U., et al., *Adsorption and photocatalytic degradation of anionic dyes on chitosan/PVA/Na–titanate/TiO2 composites synthesized by solution casting method*. Carbohydrate Polymers, 2016. **149**: p. 317–331.
[50] Mousli, F., et al., *Mixed oxide-polyaniline composite-coated woven cotton fabrics for the visible light catalyzed degradation of hazardous organic pollutants*. Cellulose, 2020. **27**(13): p. 7823–7846.
[51] Habiba, U., et al., *Degradation of methyl orange and Congo red by using chitosan/polyvinyl alcohol/TiO2 electrospun nanofibrous membrane*. International Journal of Biological Macromolecules, 2019. **131**: p. 821–827.
[52] Barzinjy, A.A., et al., *Green and eco-friendly synthesis of Nickel oxide nanoparticles and its photocatalytic activity for methyl orange degradation*. Journal of Materials Science: Materials in Electronics, 2020. **31**: p. 11303–11316.
[53] Sabir, A., T.A. Sherazi, and Q. Xu, *Porous polymer supported Ag-TiO2 as green photocatalyst for degradation of methyl orange*. Surfaces and Interfaces, 2021. **26**: p. 101318.
[54] Hao, D., et al., *Bifunctional fabric with photothermal effect and photocatalysis for highly efficient clean water generation*. ACS Sustainable Chemistry & Engineering, 2018. **6**(8): p. 10789–10797.
[55] Sionkowska, A., *Effects of solar radiation on collagen and chitosan films*. Journal of Photochemistry and Photobiology B: Biology, 2006. **82**(1): p. 9–15.
[56] Koe, W.S., et al., *An overview of photocatalytic degradation: photocatalysts, mechanisms, and development of photocatalytic membrane*. Environmental Science and Pollution Research, 2020. **27**(3): p. 2522–2565.
[57] Wu, Y., G. Lu, and S. Li, *The doping effect of Bi on TiO2 for photocatalytic hydrogen generation and photodecolorization of rhodamine B*. The Journal of Physical Chemistry C, 2009. **113**(22): p. 9950–9955.
[58] Doabi, E.H., F. Elmi, and M.M. Elmi, *Facile and novel synthesis of nitrogen doped TiO2/Acid soluble collagen-polyvinyl pyrrolidone (ASC-PVP) hybrid nanocomposite for rapid and effective photodegradation of naphthalene from water under visible light irradiation*. Journal of Photochemistry and Photobiology A: Chemistry, 2022. **425**: p. 113677.
[59] Nagaraj, S., K. Cheirmadurai, and P. Thanikaivelan, *Visible-light active collagen-TiO2 nanobio-sponge for water remediation: A sustainable approach*. Cleaner Materials, 2021. **1**: p. 100011.
[60] Chen, M.M., H.X. Li, and J.P. Lang, *Two Coordination Polymers and Their Silver (I)-Doped Species: Synthesis, Characterization, and High Catalytic Activity for the Photodegradation of Various Organic Pollutants in Water*. European Journal of Inorganic Chemistry, 2016. **2016** (15–16): p. 2508–2515.

[61] Ojha, A., et al., *An environmental approach for the photodegradation of toxic pollutants from wastewater using silver nanoparticles decorated titania-reduced graphene oxide*. Journal of Environmental Chemical Engineering, 2021: p. 105622.

[62] Wei, Y.-S., et al., *Metal–organic framework-based catalysts with single metal sites*. Chemical reviews, 2020. **120**(21): p. 12089–12174.

[63] Wang, D., D. Jana, and Y. Zhao, *Metal–organic framework derived nanozymes in biomedicine*. Accounts of Chemical Research, 2020. **53**(7): p. 1389–1400.

[64] Singh, B.K., S. Lee, and K. Na, *An overview on metal-related catalysts: metal oxides, nanoporous metals and supported metal nanoparticles on metal organic frameworks and zeolites*. Rare Metals, 2020. **39**(7): p. 751–766.

[65] Kalpana, V. and V.D. Rajeswari, *Synthesis of palladium nanoparticles via a green route using Lagenaria siceraria: assessment of their innate antidandruff, insecticidal and degradation activities*. Materials Research Express, 2018. **5**(11): p. 115406.

[66] Flores, J.G., et al., *Greener synthesis of Cu-MOF-74 and its catalytic use for the generation of vanillin*. Dalton Transactions, 2018. **47**(13): p. 4639–4645.

[67] Chen, Y., et al., *Green synthesis of Ag–TiO2 supported on porous glass with enhanced photocatalytic performance for oxidative desulfurization and removal of dyes under visible light*. ACS Sustainable Chemistry & Engineering, 2018. **6**(10): p. 13276–13286.

[68] Irfan, M., H. Munir, and H. Ismail, *Moringa oleifera gum based silver and zinc oxide nanoparticles: green synthesis, characterization and their antibacterial potential against MRSA*. Biomaterials research, 2021. **25**(1): p. 1–8.

[69] Sorbiun, M., et al., *Green synthesis of zinc oxide and copper oxide nanoparticles using aqueous extract of oak fruit hull (jaft) and comparing their photocatalytic degradation of basic violet 3*. International Journal of Environmental Research, 2018. **12**(1): p. 29–37.

[70] Saeed, M., et al., *Calotropis gigantea leaves assisted biosynthesis of ZnO and Ag@ ZnO catalysts for degradation of rhodamine B dye in aqueous medium*. Environmental Progress & Sustainable Energy, 2020. **39**(4): p. e13408.

[71] Gangwar, J. and J.K. Sebastian, *Unlocking the potential of biosynthesized zinc oxide nanoparticles for degradation of synthetic organic dyes as wastewater pollutants*. Water Science and Technology, 2021. **84**(10–11): p. 3286–3310.

[72] Oosterkamp, M.J., et al., *Nitrate and (per) chlorate reduction pathways in (per) chlorate-reducing bacteria*. 2011, Portland Press Ltd.

[73] Sánchez-Pérez, B.N., et al., *Rhizosphere and endophytic bacteria associated to Ocimum basilicum L. with decaclorobiphenyl removal potential*. Water, Air, & Soil Pollution, 2020. **231**(3): p. 1–15.

[74] Seifunnisha, O. and J. Shanthi, *Influence of Aloe vera and PEG on the evaluation of photocatalytic degradation of MG dye under UV light and visible light irradiation of ZnO nanomaterials*. Optik, 2021. **248**: p. 168064.

[75] Basavalingiah, K., et al., *Highly porous, honeycomb like Ag–ZnO nanomaterials for enhanced photocatalytic and photoluminescence studies: green synthesis using Azadirachta indica gum*. SN Applied Sciences, 2019. **1**(8): p. 1–13.

Hina Sharif, Awais Ahmad, Rafael Luque, Mabkhoot Alsaiari,
Farid A. Harraz

6 Critical approach of bionanomaterials over pharmaceutical waste

Abstract: Water pollution, particularly its harmful impacts on the environment and also on humans, has become a challenging concern all over the world. One method to overcome this emerging issue is the treatment of wastewater, but the traditional ways of treating wastewater are unable to remove the organic contaminants including emerging pharmaceutical contaminants (EPCs). Currently, the biogenic nanoparticles (bionanomaterials) have been extensively studied and found to be potent in removing EPCs from wastewater, thus protecting the environment and ecosystem. Adsorption and photocatalysis are the two advanced technologies of bionanomaterials which can be used for the effective removal of EPCs from wastewater. Even though these methods have some deficiencies and their improper discharge may also affect the environment adversely, these tenacious particles are still efficacious in terms of biomedical and environmental protecting applications. Thus, additional research is the solution of this problem to examine the safety of using these nanomaterials to the humans and environment before their use on large scale.

Keywords: Biomaterials, pharmaceutical waste, degradation, environmental safety

6.1 Introduction

Since the past few years, the extensive manufacturing and consumption of chemical products have threatened the health concerns by elevating the environmental pollution. The major risk factors contributing toward immense environmental pollution, which has gained significant global scientific attention, include pharmaceutical manufacturing and consumption, climate change, agricultural concerns, mining, urbanization and domestic activities [1]. These tenacious organic contaminants impose considerable pollution to global water cycle, as the conventional water treatment methods only partially remove these organic substances. Some of these organic contaminants escape out toward environment where they can spread to various ecological compartments [2, 3]. The emerging organic contaminants include several compounds, both metabolites and degradates (transformation products), pharmaceuticals, veterinary products, personal care products, industrial compounds, transformation products, pesticides, artificial sweeteners, methanesulfonic acid, food additives and some other nanomaterials [4, 5]. With the advancement of analytical techniques, including liquid chromatography-mass spectrometry-tandem

https://doi.org/10.1515/9783110768749-006

mass spectrometry and high-performance liquid chromatography coupled with mass spectrometry, a wide range of emerging pharmaceutical contaminants (EPCs) in wastewater have been identified and quantified which have gained considerable research among World Health Organization, the US Environmental Protection Agency, the United States Geological Survey and the European Commission (EU) [1, 6]. Using these advanced analytical techniques, the extremely low concentration of EPCs in ng/L can be estimated even in very complex matrixes in solid and liquid wastes [7]. The cognizable fatal impacts of these emerging organic contaminants in manufactured products particularly in pharmaceutical products have become an intensive subject of global public health issues because of their individual and/or synergistic effects on the earthlings and ecosystem. A huge shift of scientific research toward advanced analytical technique development, environmental monitoring systems, waste removal techniques and major risk assessments associated with the exposure of these organic contaminants reflect the critical approach of this global issue [8–10].

Although the true fate and reactions of most of the EPCs are still to be understudied, with the recent available researches and findings, the EPCs possess the potential of bioaccumulation in plant and animal tissues and persistence in the environment [5, 11]. For instance, the presence of antibiotic wastes in the water and environment has greatly increased the production of antibiotic-resistant bacteria [5, 12]. Therefore, the increased concentration of pharmaceutical wastes with the advancement in medicine system is an emerging issue in this era in the fields of engineering and environmental sciences. There is the need for time to gain effective removal of such toxic EPCs from water resources and environment [13]. The decline of freshwater resources and increased demand of water have greatly enhanced the need of water reuse projects and systems, but such systems are ineffective in removing the EPCs and other organic contaminants from wastewater. Thus, the water treatment techniques need careful and effective design and additional steps in order to ensure the water safety for consumption [14].

The advancements in the field of nanobiotechnology have offered an effective and alternative method for overcoming the environmental pollution. One way to control over this emerging global issue is the advancement in the field of nanotechnology using the process of photocatalysis or adsorption to decontaminate the water from EPCs. The processes of adsorption and photocatalysis are the two main widely employed and investigated water purification techniques, and much emphasis in research and literature is being given on the carbon-based adsorbents and/or photocatalysts for removing EPCs in the last decade [5].

6.1.1 Emerging pharmaceutical contaminants (EPCs)

Pharmaceuticals are chemical compounds, either natural or synthetic in nature, which are designed for the remedy and/or prevention of diseases and health disorders in order to increase the value to human as well as animal life [1]. These chemical compounds have diverse chemical structures, behavior, usage and metabolic fate in human and animals and also to the environment [2]. According to the pharmacological effects, pharmaceuticals include antibiotics (cephalosporins, tetracyclines and penicillins), antidiabetics (biguanides and sulfonylureas), analgesics (acetaminophen and NSAIDS), antiepileptic (carbamazepine), antianxiety (diazepam), antihistamines (famotidine and ranitidine) and antihyperlipidemics (fibric acid and clofibrate) [6]. Mompelat et al. [8] categorized the animal and human pharmaceuticals into 24 various groups, out of which 4 classes including NSAIDS, antibiotics, lipid regulators and anticonvulsants were mainly found in water. In the present days, thousands of pharmaceuticals (natural, semi-synthetic or synthetic) are commercially available for human and animal use, and the annual pharmaceutical consumption figure around the globe is about 100,000–200,000 tons. These drugs are administered via multiple of ingestion routes, and their original and partially metabolized parts are usually excreted out of the body through urine or feces due to incomplete metabolism and ultimately enter into the wastewater treatment plants [15].

Pharmaceutical wastes actually represent the pharmaceutical residues' chemical oxygen demand and biochemical oxygen demand, VOCs and pharmaceutically active compounds [16]. Based on recent researches, an immense interest due to the presence of pharmaceutical products in the form of pseudopersistent environmental contaminants has been seen due to the potential adverse and harmful effects of such EPCs on the aquatic ecosystem as well as terrestrial wildlife. Most of the pharmaceutical products are excreted out of the body unmetabolized or as degradates via urine and/or feces, and are introduced into the domestic sewerage, from where they enter into the aquatic environment [17–19]. Even though EPCs are present in water sources for more than a decade, their level in the ecosystem has started recently to be quantified as potent environmental toxicants [6, 20]. The EPCs are stable in nature and hence are enigmatic to degrade through conventional wastewater treatment plants, and thereby enter into the environment [1, 15]. The presence of pharmaceuticals and their metabolites in food and water may impose serious health impacts on both animals and humans and can also destroy the aquatic species. The most observed toxic effects of EPCs on humans include chronic toxicity, endocrine system alterations and formation of resistant bacterial strains [21]. Table 6.1 lists all the possible harmful effects of EPCs on humans and ecosystem. The EPCs have been widely identified in various environmental water sources of the USA, Holland, China, India, Iran, Spain, Canada, Germany, Italy, Brazil and South Africa [1, 12, 17, 22]. Yan et al. [17] investigated 21 EPCs in various steps of the wastewater treatment

Table 6.1: Pharmaceutical drugs and effects of their EPCs.

Pharmaceutical class	Examples	Pharmacological effects	Harmful effects of their EPCs	References
Analgesics	NSAIDS (ibuprofen, diclofenac, naproxen), paracetamol and phenazopyridine	Reduces pain, anti-inflammatory and antipyretic	Diclofenac is found to be bioaccumulated. Analgesic can induce negative effects on development	[71, 72]
Antibiotics	Amoxicillin, cefixime, ciprofloxacin, ofloxacin, metronidazole, sulfonamides and tetracyclines	Kill microbes or inhibit their growth	Development of antibiotic-resistant bacteria and impair the plant growth when present in soil. Sulfonamides have been reported to cause birth defects	[73, 74]
Anticoagulants	Warfarin	Inhibit the formation of blood clot	Its prolonged exposure can cause bleeding due to persistent inhibition of vitamin K	[75]
Antidiabetics	Acarbose, pramlintide, metformin and insulin	Reduce hyperglycemia	Metformin is a reported endocrine-disrupting chemical (EDC) and is highly mobile in the ecosystem and is not degraded easily	[76]
Antihistamines	Diphenhydramine	Used for the treatment of allergic reactions	This EPC can cause acute and chronic toxicity to many aquatic species	[77]
Antipsychotics	Clozapine, chlorpromazine, olanzapine. loxapine and risperidone	Treatment of mental disorders	These EPCs have persistent and bioaccumulative nature and have been reported to be toxic to human health and ecosystem	[4, 78]
Beta-blocking agents	Propranolol and metoprolol	Antihypertensive	Toxic to aquatic species, zooplankton and phytoplankton	[66]
Fibrates	Gemfibrozil	Antihyperlipidemic	Toxic to aquatic species	[71]

plants in the Southwest China. The nature of most of the pharmaceuticals in the environment is also complex and diverse but unfortunately there is only limited knowledge and research in this regard. Some EPCs are water soluble and break down easily due to their hydrophilic behavior and hence have short life span; whereas some can maintain their life span for more than 1 year without degradation such as sulfamethoxazole and naproxen, and some can even survive for several years such as clofibric acid [4]. The sources of entrance of pharmaceutical products in wastewater include domestic sewage, landfill sites, septic tanks, industrial effluents, agricultural practices, urban sewerage water, showering and bathing, and in this way these products enter the water cycle [23].

Others routes of entering these contaminants in the environment include human excretion, recreational activities and direct discharge of such products through pharmaceutical-producing plants [1, 21]. The most of EPCs found in aquatic system are the drug of abuse due to their psychoactive properties [24]. Inappropriate disposal of expired or unused drugs in the toilets or landfills and/or pharmaceutical manufacturing residues which are thrown directly into the sewerage can also be regarded as a significant risk factor of potential contamination [25]. Baker and Kasprzyk-Hordern [15] investigated the spatial and temporal presence of more than 60 EPCs including illicit drugs and their metabolites in both wastewaters and a major river in the United Kingdom. Chen et al. [22] investigated the presence of about 43 emerging environmental contaminants including 34 pharmaceuticals and 9 endocrine-disrupting chemicals in both wastewater and groundwater of North China.

6.1.2 Fate of EPCs in the environment

The expanding accumulation of EPCs in the environment has greatly gained the attention of public and scientists in the last 15 years. Based on recent several researches and investigations, these environmental toxicants are continuously identified and detected in various water sources, especially in drinking water. Hence, the fate and ecotoxicological effects of these EPCs should be studied after their escape from the conventional wastewater treatment plants. The major factors that govern the fate of such organic compounds include partition solubility index, coefficient and other environmental factors such as pH, temperature, humus content and salinity [26]. The possible processes of EPCs in treatment plants are diverse: (a) harmless inorganic compounds are converted into water and carbon dioxide through chemical, biological, physicochemical transformation; (b) the lipophilic nature of innocuous EPCs may be partially adsorbed on the sediments or sludge; and (c) ultimately such products are discharged as bulk or in the form of conjugate products into the receiving water [27]. The mobile nature of EPCs in the environment and their retention behavior on to the sediments depend upon their lipophilicity. Some pharmaceutical products such as triclosan, diclofenac, fluoxetine, fenofibrate, diazepam, iodoarene and albuterol undergo significant photodegradation, while tetracyclines and phthalates remain in the environment [28]. Mompelat et al. [8] reported the presence of acridine in the ecosystem, a photodecomposed product of carbamazepine, and this waste compound is found to be highly toxic and carcinogenic. Tetracyclines, the most commonly used antibiotics, also impose serious human health impacts and ecological risks because about 70% of tetracyclines are excreted out of the body unchanged, and their hydrophilic and less volatile nature are the causes for their persistent presence in the ecosystem [21]. The residues of tetracyclines in the environment accelerates the formation of antibiotic-resistant microstrains which then ultimately cause serious health impacts to humans through drinking water or even irrigation water used for crops. These antimicrobial agents possess the potential to disturb the normal flora of humans and also elevate the risk of several infections [29, 30]. Moreover, the presence of oxytetracycline and chlortetracycline also impose some estrogenic adverse effects such as hindering the first day of the reproduction, the number of siblings per female and the growth rate of populations [21, 31].

6.1.3 Technological advancements for the wastewater treatment plants

Wastewater can be treated using several ways, such as unplanned, direct or indirect. The simplest water treatment technique is the unplanned approach in which treated water is directly released into the natural water bodies [32], whereas the direct and indirect wastewater treatment techniques include some additional advanced steps of water treatment before that water is released into the environment, such as membrane filtration, chlorination, ion exchange, carbon adsorption and some advanced oxidation processes including UV radiation and/or ozone [33]. For the proper removal of EPCs from wastewater, the selection of proper combination of advanced tertiary steps of water treatment is very important [34]. The direct water treatment plants feed the treated water into the drinking water treatment plants after their tertiary steps while the indirect potable reuse plant systems release the treated water into the natural water systems including river, groundwater aquifer or sea [5].

6.1.4 Critical approach of bionanomaterials for wastewater treatment

The use of nanomaterials has been widely increased since the last decade and has become an extensive part of our daily used products. The global market of these nanomaterials had crossed US $6.2 billion in 2015 [35]. The nanomaterials possess huge productivity and efficiency in optical, catalysis, thermal, electrical and mechanical conductivity as well as biological and chemical activity due to their extra ordinary porosity and high surface-area-to-volume ration. Moreover, these effective compounds can be easily tailored to the required specification and uses [36, 37]. The biosynthesis of nanomaterials, also known as green synthesis, uses redox process in which the organisms and/or their metabolites are used to reduce the metal ions into stable nanomaterials through nicotinamide adenine dinucleotide phosphate hydrogen-dependent reductase enzymes. Such types of nanomaterials which are produced through biological pathways are better known as biogenic nanomaterials or biomaterials. They possess high sensitivity and biocompatibility and also have prominent applications in agriculture, biomedical, environmental remediation and electronic sectors [38, 39]. The green synthesis of metal oxide nanoparticles possesses unique properties and therefore gained significant attention. The production of nanomaterials through plant extracts and microorganisms has gained significant attention due to their simple process, environmental friendliness, economic feasibility and ease of scale-up [40, 41]. The use and synthesis of bionanomaterials show a rapidly growing interest in both research and development due to their extra ordinary properties [36, 42].

The two advanced technologies of bionanomaterials such as adsorption and photocatalysis can be employed effectively for the removal of EPCs from wastewater. Graphene-based and metal-based nanoparticles have shown promising water treatment techniques due to their extra ordinary and patronizing properties of removing and breaking down of EPCs [5].

6.1.5 Adsorption using bionanomaterials

Adsorption is a surface phenomenon in which organic and inorganic compounds are attached or adhered to the surface of adsorbents due to physical and chemical interactions caused by electrostatic interactions and/or van der Waals forces. The qualities of an effective adsorbent to be used for the removal of EPCs from water include inert, resistant and high adsorption capability. The factors that govern the process of adsorption include pH, temperature, concentration of contaminants and their particle size, contact time and physical and chemical nature of the adsorbents [43]. The EPCs having more aromatic rings in their structures have swift adsorption rates making their removal easy from the wastewater [44]. In graphene-based nanomaterials, the process of adsorption is fastest due to π–π and hydrophobic interactions between the aromatic rings, electrostatic interactions and hydrogen bonding due to the presence of COOH, NH_2 and OH groups [45, 46].

The use of nanotechnology at industrial scale assures a guaranteed removal of EPCs in water treatment process. Activated carbon is an industrially effective adsorbent, while other adsorbents which can credibly remove EPCs include graphene, clay minerals, carbon nanotubes, polymeric materials and siliceous adsorbents. Among all of these adsorbents, graphene-based nanomaterials are the good candidates for the adsorption of EPCs. Due to their remarkable properties and unique structures, they exhibit markedly rapid and effective adsorption kinetics. Their large surface-area-to-volume ratio, physicochemical properties and electrostatic interactions with the contaminants make them a perfect choice as EPC adsorbents [47–50]. Moreover, the conjugated π-region of graphene makes it favorable for the removal of both organic and inorganic contaminants [51, 52].

To increase the adsorption capacity and reusability of graphene, different attempts have been done to modify its surface. For instance, the modification of graphene oxide (GO) with ferrosoferric oxide (Fe_3O_4), manganese oxide (MnO_2) and Fe/Cu and formation of graphene hydrogels show however low surface area but large adsorption capability as compared to the unmodified graphene [5, 48, 50]. In the same way, the activated graphene exhibits large surface area showing high adsorption capability toward ciprofloxacin as compared to the magnetic chitosan-grafted GO composite [5]. Table 6.2 represents the list of adsorbents which are made by

Table 6.2: List of adsorbents used for EPC removal.

Pharmaceuticals	EPCs	Adsorbents	Removal (%)	References
Antibiotic	Amoxicillin, cephalexin, ciprofloxacin, norfloxacin, ofloxacin, sulfadiazine, tetracycline and oxytetracycline	Graphene, activated graphene, graphene oxide, graphene hydrogel, magnetic graphene nanoplatelets, reduced graphene oxide and magnetic chitosan-grafted graphene oxide composite	75–100%	[79–85]
Analgesics	Acetaminophen, aspirin, diclofenac sodium, ibuprofen, naproxen and salicylic acid	Graphene, graphene oxide, reduced graphene oxide, graphene oxide-based hydrogels, GO/β-Bi_2O_3/TiO_2/$Bi_2Ti_2O_7$ heterojuncted nanocomposite, metal organic frameworks and functionalized graphene	80–99%	[86–91]
Antihistamines	Diphenhydramine	Reduced GO-TiO_2 composites	98–100%	[92]
Antidiabetic	Metformin	GO	80%	[93]
Anticonvulsant	Carbamazepine	Graphene, TiO_2–reduced GO, GO/β-Bi_2O_3/TiO_2/$Bi_2Ti_2O_7$ heterojuncted nanocomposite	97–100%	[87, 88, 94]

modifying the graphene's surface found to be reported for EPC removal. The use of nanomaterials makes the adsorption process fast and accurate for removing EPCs from the aquatic ecosystems. Some adsorbents ensure 50% removal of EPCs, while several adsorbents ensure an efficiency of more than 99%. Nevertheless, there are some disadvantages of adsorbents as well: EPCs limit the adsorbent reusability, EPCs may create a potential novel environmental contaminants and the interaction of EPCs with adsorbents is not a permanent reaction; therefore, some additional steps must be included in the water treatment plants [5]. Other problems associated with the use of GO and its nanomaterials include high cost, exhausting and time-consuming process which limit their use [53]. Due to the rapid growth and advancements in the field of science and technology, the material reusability issues may be resolved in near future which then can increase potential usage of GO and its nanomaterials at the commercial level [5].

6.2 Photocatalysis

Photocatalysts actually refer to the semiconducting materials that are metal oxides (zinc oxide and titanium oxide) which possess he potential of absorbing light and generating reactive oxygen species ultimately degrading the pollutants [54 55]. Composite composing of metal-doped semiconductors and appropriate adsorbents can be employed for a photocatalytic process [55]. Several semiconductor nanomaterials such as TiO_2, ZnO, Fe_2O_3, CdS and CeO_2 have been used successfully as photocatalysts [56]. Table 6.3 shows the list of photocatalysts used to degrade EPCs, and some of these photocatalysts are modified using graphene and graphene-based materials for enhancing the capability of adsorption and degradation. The use of nanomaterials as photocatalysts has offered a staggering, surrogate and eco-friendly method for the removal of EPCs from wastewater [57].

TiO_2 is one of the photocatalysts which has been extensively used by several researchers and is found to be potent for the effective degradation of several EPCs such as phthalic acid, benzoic acid, *o*-nitrobenzoic acid, *o*-chlorobenzoic acid and nitrobenzene [56]. The source of light which is used for photocatalysis causes the excitation of outmost electrons creating an electron–hole pair that is responsible for the redox reaction, ultimately degrading the pollutants. However, the process of EPC degradation due to photocatalysis is still complex but it can be thought that the electrons may form the superoxide anion, and hole may cause reduction processes forming hydroxyl radicals. The superoxide anion and hydroxyl radicals which are formed in this process are the cause of degradation of certain compounds [5]. This process of photodegradation depends upon several factors such as EPC structures, concentration, experimental conditions, water chemistry and nanomaterial loading [58]. In the photocatalytic degradation of acetaminophen, the hydroxyl

Table 6.3: Photocatalysts used for the degradation of EPCs.

Pharmaceuticals	EPCs	Photocatalysts	Source of reaction	Removal of EPCs (%)	References
Analgesics	Diclofenac, paracetamol, ibuprofen, phenazopyridine and naproxen	Graphene/TiO_2 nanotubes, ZnO/sepiolite, carbon-doped TiO_2, TiO_2/montmorillonite, TiO_2/cellulosic fiber, carbon-doped TiO_2 on zeolites, TiO_2 rutile nanorods, ZnO–TiO_2 and TiO_2–P25 nanoparticles in photoreactors	Solar light and/or UV–visible light	75–100%	[44, 66, 89, 94–98]
Antibiotics	4-Chlorophenol, cefixime, ciprofloxacin, levofloxacin, metronidazole, oxytetracycline, penicillin G, sulfadiazine and trimethoprim	ZrO_2/Fe_3O_4/chitosan, nano-N-TiO_2/GO/titan grid sheets, TiO_2 nanocomposites, Bi_2WO_6 nanocuboids, TiO_2/montmorillonite, graphene/TiO_2/ZSM-5 composites, r-TiO_2 nanocatalysts and zeolite coated with TiO_2	Visible and UV	65–95%	[99–106]
Anticoagulant	Warfarin	PVDF membrane with TiO_2	Lower pressure UV light	–	[67]
Anticonvulsant	Carbamazepine	BiOCl microsphere and TiO_2–rGO	Visible light and high-pressure UV light	70%	[65, 94]
Antipsychotics	Loxapine	TiO_2	Solar spectrum	99.8%	[107]
Beta-blocking agent	Metoprolol	PVDF membrane with TiO_2	Low-pressure UV light	–	[67]

radical generated is responsible for hydroxylation and then subsequent breakdown of aromatic rings [44]. Based on recent research, it has been observed that the increased concentration of superoxide anion than hydroxyl ion can also degrade acetaminophen by acting as a Lewis acid [59].

Although the process of photocatalysis is offering several benefits over traditional wastewater treatment techniques, there are some limitations as well, which should be overcome to increase effectiveness of this process. This process needs energy required for electron excitation which ultimately enhances the cost requirement. Moreover, charge carrier transfer rate, recombination rate and charge carrier travel time are also some of the limitations of photocatalysis [60]. Structural alterations and/or doping are the ways of material alterations which may improve the process efficiency. Both these methods can decrease the bandgap ultimately reducing the recombination rates. The application of plasmonic agents can also decrease the energy requirements [61, 62]. The probable consequences of photocatalytic materials on the environment are their another limitation. Their potential transformation products can have serious impacts if they are released into the environs, as they are manifested with the harmful effects of phenol derivatives released after the photocatalytic degradation of diclofenac [63]. Pharmaceutical products often contain aromatic rings which can be converted into phenolic compounds after their disposal if not degraded properly, and phenolic compounds are the toxic agents. Except that, the pharmaceutical waste products may form acids (e.g., paracetamol degradation) and such acids also impose adverse impacts on the environs [59, 63]. The instability of photocatalysts in the water is another concern which in turn can affect the environment, thereby the mechanism of degradation through photocatalysts should be understood before their usage for wastewater treatment [64]. To overcome these limitations, the use of stabilizing agents and generation of composites can improve the efficiency and stability of photocatalysts, and their harmful impacts on the environment can be overcome. In spite of these limitations, photocatalysts offer promising applications at commercial scale, as many of the photocatalysts are effective in degrading EPCs using low-power UV and/or visible light [5]. Regardless of use of light source, photocatalysts can lower the energy requirements for water decontamination. Many of the nonbiodegradable EPCs such as iopromide, norfloxacin and carbamazepine have been removed using photocatalysts [65–67]. However, additional research is still needed to examine the safety of using these agents to the environment and humans before their use on large scale [5].

Currently, several materials namely optical fiber-coated materials and magnetic nanocomposites have been developed to enhance the uses of photocatalysts. The degradation property of magnetic composites makes their use effective for the easier removal of photocatalysts from the water, ultimately reducing their harmful impact on the environment [68, 69]. Moreover, development of photocatalytic materials on optical fibers has made them immobilized, ultimately increasing their usage [70].

6.3 Conclusion

The green synthesis and utilization of bionanomaterials offer a promptly growing area in extensive research and environmental protection from hazardous compounds in terms of their eco-friendly and efficient properties. The use of these bionanomaterials offers an indirect and alternative process for the removal of EPCs from wastewater through adsorption and/or photocatalysis, but these technologies require thorough risk assessments and safety evaluation for their proper usage and disposal. There is no legislation which pertains to the maximum allowable concentration of EPCs in the wastewater, which should be regulated to ensure the environmental safety and human health. Hence, further research and investigations are needed for maximum EPC removal from wastewater and employment of nanotechnology for wastewater treatment.

References

[1] Tijani, J.O., et al., *Pharmaceuticals, endocrine disruptors, personal care products, nanomaterials and perfluorinated pollutants: a review.* Environ Chem Lett, 2016. **14**(1): p. 27–49.

[2] Fawell, J. and C.N. Ong, *Emerging contaminants and the implications for drinking water.* Int J Water Resour Dev, 2012. **28**(2): p. 247–263.

[3] Miraji, H., et al., *Research trends in emerging contaminants on the aquatic environments of Tanzania.* Scientifica (Cairo), 2016. **2016**: p. 3769690.

[4] Fent, K., A.A. Weston, and D. Caminada, *Ecotoxicology of human pharmaceuticals.* Aquat Toxicol, 2006. **76**(2): p. 122–159.

[5] Fanourakis, S.K., et al., *Nano-based adsorbent and photocatalyst use for pharmaceutical contaminant removal during indirect potable water reuse.* NPJ Clean Water, 2020. **3**(1): p. 1.

[6] Rivera-Utrilla, J., et al., *Pharmaceuticals as emerging contaminants and their removal from water. A review.* Chemosphere, 2013. **93**(7): p. 1268–1287.

[7] Petrović, M., et al., *Liquid chromatography–tandem mass spectrometry for the analysis of pharmaceutical residues in environmental samples: a review.* J Chromatogr A, 2005. **1067**(1): p. 1–14.

[8] Mompelat, S., B. Le Bot, and O. Thomas, *Occurrence and fate of pharmaceutical products and by-products, from resource to drinking water.* Environ Int, 2009. **35**(5): p. 803–814.

[9] Jiang, J.-Q., Z. Zhou, and V.K. Sharma, *Occurrence, transportation, monitoring and treatment of emerging micro-pollutants in waste water – a review from global views.* Microchem J, 2013. **110**: p. 292–300.

[10] Riva, F., et al., *Risk assessment of a mixture of emerging contaminants in surface water in a highly urbanized area in Italy.* J Hazard Mater, 2019. **361**: p. 103–110.

[11] Xiang, J., et al., *The fate and risk assessment of psychiatric pharmaceuticals from psychiatric hospital effluent.* Ecotoxicol Environ Saf, 2018. **150**: p. 289–296.

[12] Mirzaei, R., et al., *Antibiotics in urban wastewater and rivers of Tehran, Iran: Consumption, mass load, occurrence, and ecological risk.* Chemosphere, 2019. **221**: p. 55–66.

[13] Khetan, S.K. and T.J. Collins, *Human pharmaceuticals in the aquatic environment: a challenge to green chemistry*. Chem Rev, 2007. **107**(6): p. 2319–2364.

[14] Wintgens, T., et al., *Emerging Contaminants and Treatment Options in Water Recycling for Indirect Potable Use*. Water Sci Technol, 2008. **57**: p. 99–107.

[15] Baker, D.R. and B. Kasprzyk-Hordern, *Spatial and temporal occurrence of pharmaceuticals and illicit drugs in the aqueous environment and during wastewater treatment: New developments*. Sci Total Environ, 2013. **454–455**: p. 442–456.

[16] Ma, D., et al., *Critical review of advanced oxidation processes in organic wastewater treatment*. Chemosphere, 2021. **275**: p. 130104.

[17] Yan, Q., et al., *Occurrence and fate of pharmaceutically active compounds in the largest municipal wastewater treatment plant in Southwest China: Mass balance analysis and consumption back-calculated model*. Chemosphere, 2014. **99**: p. 160–170.

[18] Leung, H.W., et al., *Distribution, fate and risk assessment of antibiotics in sewage treatment plants in Hong Kong, South China*. Environment International, 2012. **42**: p. 1–9.

[19] M., J.L., *Antibiotics and Antibiotic Resistance Genes in Natural Environments*. Science, 2008. **321**(5887): p. 365–367.

[20] Kolpin, D.W., et al., *Pharmaceuticals, Hormones, and Other Organic Wastewater Contaminants in U.S. Streams, 1999–2000: A National Reconnaissance*. Environ Sci Technol, 2002. **36**(6): p. 1202–1211.

[21] Daghrir, R. and P. Drogui, *Tetracycline antibiotics in the environment: a review*. Environmental Chemistry Letters, 2013. **11**(3): p. 209–227.

[22] Chen, F., et al., *Distribution and accumulation of endocrine-disrupting chemicals and pharmaceuticals in wastewater irrigated soils in Hebei, China*. Environ Pollut, 2011. **159**(6): p. 1490–1498.

[23] Rodil, R., et al., *Emerging pollutants in sewage, surface and drinking water in Galicia (NW Spain)*. Chemosphere, 2012. **86**(10): p. 1040–1049.

[24] Castiglioni, S., E. Zuccato, and R. Fanelli, *Illicit drugs in the environment: occurrence, analysis, and fate using mass spectrometry*. 2011, Oxford: Wiley-Blackwell.

[25] Reddersen, K., T. Heberer, and U. Dünnbier, *Identification and significance of phenazone drugs and their metabolites in ground- and drinking water*. Chemosphere, 2002. **49**(6): p. 539–44.

[26] Lapworth, D.J., et al., *Emerging organic contaminants in groundwater: A review of sources, fate and occurrence*. Environ Pollut, 2012. **163**: p. 287–303.

[27] Richardson, S.D., *Environmental Mass Spectrometry: Emerging Contaminants and Current Issues*. Anal Chem, 2012. **84**(2): p. 747–778.

[28] Padhye, L.P., et al., *Year-long evaluation on the occurrence and fate of pharmaceuticals, personal care products, and endocrine disrupting chemicals in an urban drinking water treatment plant*. Water Research, 2014. **51**: p. 266–276.

[29] Stackelberg, P.E., et al., *Efficiency of conventional drinking-water-treatment processes in removal of pharmaceuticals and other organic compounds*. Sci Total Environ, 2007. **377**(2): p. 255–272.

[30] Arikan, O.A., C. Rice, and E. Codling, *Occurrence of antibiotics and hormones in a major agricultural watershed*. Desalination, 2008. **226**(1): p. 121–133.

[31] Park, S. and K. Choi, *Hazard assessment of commonly used agricultural antibiotics on aquatic ecosystems*. Ecotoxicology, 2008. **17**(6): p. 526–38.

[32] Roccaro, P., *Treatment processes for municipal wastewater reclamation: The challenges of emerging contaminants and direct potable reuse*. Curr Opin Environ Sci Health, 2018. **2**.

[33] Rojas, M.R., et al., *Assessment of the Effectiveness of Secondary Wastewater Treatment Technologies to Remove Trace Chemicals of Emerging Concern*. Crit Rev Environ Sci Technol, 2013. **43**(12): p. 1281–1314.

[34] Heberer, T., K. Reddersen, and A. Mechlinski, *From municipal sewage to drinking water: fate and removal of pharmaceutical residues in the aquatic environment in urban areas*. Water Sci Technol, 2002. **46**(3): p. 81–8.

[35] Kunhikrishnan, A., et al., *Sources, Distribution, Environmental Fate, and Ecological Effects of Nanomaterials in Wastewater Streams*. Crit Rev Environ Sci Technol, 2015. **45**(4): p. 277–318.

[36] Ahmed, S.F., et al., *Green approaches in synthesising nanomaterials for environmental nanobioremediation: Technological advancements, applications, benefits and challenges*. Environ Res, 2022. **204**: p. 111967.

[37] Velu, R., et al., *A comprehensive review on bio-nanomaterials for medical implants and feasibility studies on fabrication of such implants by additive manufacturing technique*. Materials, 2019. **13**(1): p. 92.

[38] Gour, A. and N.K. Jain, *Advances in green synthesis of nanoparticles*. Artif Cells Nanomed Biotechnol, 2019. **47**(1): p. 844–851.

[39] Rana, A., K. Yadav, and S. Jagadevan, *A comprehensive review on green synthesis of nature-inspired metal nanoparticles: Mechanism, application and toxicity*. J Clean Prod, 2020. **272**: p. 122880.

[40] Rostami-Vartooni, A., M. Alizadeh, and M. Bagherzadeh, *Green synthesis, characterization and catalytic activity of natural bentonite-supported copper nanoparticles for the solvent-free synthesis of 1-substituted 1H-1,2,3,4-tetrazoles and reduction of 4-nitrophenol*. Beilstein J Nanotechnol, 2015. **6**: p. 2300–9.

[41] de Jesus Oliveira, A.C., et al., *Solvent-free production of phthalated cashew gum for green synthesis of antimicrobial silver nanoparticles*. Carbohydr Polym, 2019. **213**: p. 176–183.

[42] Ijaz, I., et al., *Detail review on chemical, physical and green synthesis, classification, characterizations and applications of nanoparticles*. Green Chem Lett Rev, 2020. **13**(3): p. 223–245.

[43] Guedidi, H., et al., *The effects of the surface oxidation of activated carbon, the solution pH and the temperature on adsorption of ibuprofen*. Carbon, 2013. **54**: p. 432–443.

[44] Yang, L., L.E. Yu, and M.B. Ray, *Degradation of paracetamol in aqueous solutions by TiO2 photocatalysis*. Water Res, 2008. **42**(13): p. 3480–3488.

[45] Zhu, D. and J.J. Pignatello, *Characterization of aromatic compound sorptive interactions with black carbon (charcoal) assisted by graphite as a model*. Environ Sci Technol, 2005. **39**(7): p. 2033–2041.

[46] Cortés-Arriagada, D., L. Sanhueza, and K. Wrighton-Araneda, *Removal of 4-chlorophenol using graphene, graphene oxide, and a-doped graphene (A = N, B): A computational study*. Int J Quantum Chem, 2013. **113**: p. 1931–1939.

[47] Zhu, Y., et al., *Graphene and graphene oxide: synthesis, properties, and applications*. Adv Mater, 2010. **22**(35): p. 3906–3924.

[48] Soldano, C., A. Mahmood, and E. Dujardin, *Production, properties and potential of graphene*. Carbon, 2010. **48**(8): p. 2127–2150.

[49] Kyzas, G., E. Deliyanni, and K. Matis, *Graphene oxide and its application as an adsorbent for wastewater treatment*. J Chem Technol Biotechnol, 2014. **89**: p. 192–205.

[50] Yusuf, M., et al., *Applications of graphene and its derivatives as an adsorbent for heavy metal and dye removal: a systematic and comprehensive overview*. RSC Advances, 2015. **5**: p. 50392–50420.

[51] Al-Khateeb, L.A., S. Almotiry, and M.A. Salam, *Adsorption of pharmaceutical pollutants onto graphene nanoplatelets*. Chem Eng J, 2014. **248**: p. 191–199.

[52] Singh, S., et al., *Adsorption and detoxification of pharmaceutical compounds from wastewater using nanomaterials: A review on mechanism, kinetics, valorization and circular economy*. J Environ Manage, 2021. **300**: p. 113569.

[53] Kyzas, G.Z., et al., *Removal of dorzolamide from biomedical wastewaters with adsorption onto graphite oxide/poly(acrylic acid) grafted chitosan nanocomposite*. Bioresour Technol, 2014. **152**: p. 399–406.

[54] Fernández-Castro, P., et al., *Insight on the fundamentals of advanced oxidation processes. Role and review of the determination methods of reactive oxygen species*. J Chem Technol Biotechnol, 2015. **90**(5): p. 796–820.

[55] Nosaka, Y. and A.Y. Nosaka, *Generation and Detection of Reactive Oxygen Species in Photocatalysis*. Chem Rev, 2017. **117**(17): p. 11302–11336.

[56] Varma, K.S., et al., *Photocatalytic degradation of pharmaceutical and pesticide compounds (PPCs) using doped TiO2 nanomaterials: A review*. Water-Energy Nexus, 2020. **3**: p. 46–61.

[57] Singh, A., et al., *Green synthesis of metallic nanoparticles as effective alternatives to treat antibiotics resistant bacterial infections: A review*. Biotechnol Rep, 2020. **25**: p. e00427.

[58] Yang, H., et al., *Photocatalytic degradation kinetics and mechanism of environmental pharmaceuticals in aqueous suspension of TiO2: A case of sulfa drugs*. Catalysis Today, 2010. **153**(3): p. 200–207.

[59] Yang, L., L.E. Yu, and M.B. Ray, *Photocatalytic Oxidation of Paracetamol: Dominant Reactants, Intermediates, and Reaction Mechanisms*. Environ Sci Technol, 2009. **43**(2): p. 460–465.

[60] Zhang, T., et al., *Nanotechnologies for water environment applications*. 2009. 1–630.

[61] Dong, H., et al., *An overview on limitations of TiO2-based particles for photocatalytic degradation of organic pollutants and the corresponding countermeasures*. Water Res, 2015. **79**: p. 128–146.

[62] Alvarez, P.J.J., et al., *Emerging opportunities for nanotechnology to enhance water security*. Nat Nanotechnol, 2018. **13**(8): p. 634–641.

[63] Banaschik, R., et al., *Degradation and intermediates of diclofenac as instructive example for decomposition of recalcitrant pharmaceuticals by hydroxyl radicals generated with pulsed corona plasma in water*. J Hazard Mater, 2018. **342**: p. 651–660.

[64] Misra, S.K., et al., *The complexity of nanoparticle dissolution and its importance in nanotoxicological studies*. Sci Total Environ, 2012. **438**: p. 225–232.

[65] Gao, X., et al., *Highly efficient and visible-light-driven BiOCl for photocatalytic degradation of carbamazepine*. J Alloys Compd, 2018. **757**: p. 455–465.

[66] An, Y., et al., *Adsorption and photocatalytic degradation of pharmaceuticals and pesticides by carbon doped-TiO2 coated on zeolites under solar light irradiation*. Water Sci Technol, 2016. **73**(12): p. 2868–81.

[67] Paredes, L., et al., *Application of immobilized TiO2 on PVDF dual layer hollow fibre membrane to improve the photocatalytic removal of pharmaceuticals in different water matrices*. Appl Catal B 2019. **240**: p. 9–18.

[68] Li, Q., et al., *Photo-Fenton degradation of amoxicillin via magnetic TiO2-graphene oxide-Fe3O4 composite with a submerged magnetic separation membrane photocatalytic reactor (SMSMPR)*. J Hazard Mater, 2019. **373**: p. 437–446.

[69] Sun, Q., et al., *Synergistic operation of photocatalytic degradation and Fenton process by magnetic Fe3O4 loaded TiO2*. Appl Surf Sci, 2018. **430**: p. 399–406.

[70] O'Neal Tugaoen, H., et al., *Compact light-emitting diode optical fiber immobilized TiO2 reactor for photocatalytic water treatment*. Sci Total Environ, 2018. **613–614**: p. 1331–1338.

[71] Bruce, G.M., R.C. Pleus, and S.A. Snyder, *Toxicological Relevance of Pharmaceuticals in Drinking Water*. Environ Sci Technol, 2010. **44**(14): p. 5619–5626.

[72] Escher, B.I., et al., *In vitro assessment of modes of toxic action of pharmaceuticals in aquatic life*. Environ Sci Technol, 2005. **39**(9): p. 3090–100.
[73] Baquero, F., J.L. Martínez, and R. Cantón, *Antibiotics and antibiotic resistance in water environments*. Curr Opin Biotechnol, 2008. **19**(3): p. 260–5.
[74] Crider, K.S., et al., *Antibacterial medication use during pregnancy and risk of birth defects: National Birth Defects Prevention Study*. Arch Pediatr Adolesc Med, 2009. **163**(11): p. 978–85.
[75] Organization, W.H., *Warfarin: health and safety guide*. 1995.
[76] Lopez, B., et al., *Screening of French groundwater for regulated and emerging contaminants*. Sci Total Environ, 2015. **518–519**: p. 562–73.
[77] Du, B., et al., *Comparison of contaminants of emerging concern removal, discharge, and water quality hazards among centralized and on-site wastewater treatment system effluents receiving common wastewater influent*. Sci Total Environ, 2014. **466–467**: p. 976–984.
[78] David, A., et al., *Concentrating mixtures of neuroactive pharmaceuticals and altered neurotransmitter levels in the brain of fish exposed to a wastewater effluent*. Sci Total Environ, 2018. **621**: p. 782–790.
[79] Wu, S., et al., *Adsorption of ciprofloxacin onto biocomposite fibers of graphene oxide/calcium alginate*. Chem Eng J, 2013. **230**: p. 389–395.
[80] Yang, G.C. and P.L. Tang, *Removal of phthalates and pharmaceuticals from municipal wastewater by graphene adsorption process*. Water Sci Technol, 2016. **73**(9): p. 2268–74.
[81] Wang, F., et al., *Removal of ciprofloxacin from aqueous solution by a magnetic chitosan grafted graphene oxide composite*. J Mol Liq, 2016. **222**: p. 188–194.
[82] Yu, F., J. Ma, and D. Bi, *Enhanced adsorptive removal of selected pharmaceutical antibiotics from aqueous solution by activated graphene*. Environ Sci Pollut Res Int, 2015. **22**(6): p. 4715–24.
[83] Ma, J., et al., *Water-enhanced Removal of Ciprofloxacin from Water by Porous Graphene Hydrogel*. Sci Rep, 2015. **5**(1): p. 13578.
[84] Kerkez, Ö., Ş. Bayazit, and M. Salam, *Antibiotic Amoxicillin removal from aqueous solution using magnetically modified graphene nanoplatelets*. J Ind Eng Chem 2016. **36**.
[85] Chen, H., B. Gao, and H. Li, *Removal of sulfamethoxazole and ciprofloxacin from aqueous solutions by graphene oxide*. J Hazard Mater, 2015. **282**: p. 201–207.
[86] Tao, H., et al., *Enhanced photoactivity of graphene/titanium dioxide nanotubes for removal of Acetaminophen*. Appl Surf Sci, 2014. **324**.
[87] Lee, G., et al., *Fabrication of graphene-oxide/β-Bi2O3/TiO2/Bi2Ti2O7 heterojuncted nanocomposite and its sonocatalytic degradation for selected pharmaceuticals*. Chemosphere, 2018. **212**: p. 723–733.
[88] Rizzo, L., et al., *Advanced treatment of urban wastewater by sand filtration and graphene adsorption for wastewater reuse: effect on a mixture of pharmaceuticals and toxicity*. J Environ Chem Eng, 2015. **3**(1): p. 122–128.
[89] Hiew, B.Y.Z., et al., *Adsorptive removal of diclofenac by graphene oxide: Optimization, equilibrium, kinetic and thermodynamic studies*. J Taiwan Inst Chem Eng 2019. **98**: p. 150–162.
[90] Seo, P.W., et al., *Adsorptive removal of pharmaceuticals and personal care products from water with functionalized metal-organic frameworks: remarkable adsorbents with hydrogen-bonding abilities*. Sci Rep, 2016. **6**(1): p. 34462.
[91] Xin Jiat, L., N. Chemmangattuvalappil, and L.Y. Lee, *Adsorptive removal of salicylic acid from aqueous solutions using new graphene-based nanosorbents*. Chem Eng Trans, 2015. **45**: p. 1387–1392.

[92] Pastrana-Martínez, L.M., et al., *Advanced nanostructured photocatalysts based on reduced graphene oxide–TiO2 composites for degradation of diphenhydramine pharmaceutical and methyl orange dye*. Appl Catal B, 2012. **123–124**: p. 241–256.

[93] Zhu, S., et al., *Adsorption of emerging contaminant metformin using graphene oxide*. Chemosphere, 2017. **179**: p. 20–28.

[94] Lin, L., H. Wang, and P. Xu, *Immobilized TiO2-reduced Graphene Oxide Nanocomposites on Optical Fibers as High Performance Photocatalysts for Degradation of Pharmaceuticals*. Chem Eng J, 2016. **310**.

[95] de Luna, M.D.G., et al., *Photocatalytic oxidation of acetaminophen using carbon self-doped titanium dioxide*. Sustain Environ Res, 2016. **26**(4): p. 161–167.

[96] Huyen, T.T.T., et al., *Enhanced photocatalytic activity of {110}-faceted TiO_2 rutile nanorods in the photodegradation of hazardous pharmaceuticals*. Nanomaterials, 2018. **8**(5): p. 276.

[97] Shargh, M. and M.A. Behnajady, *A high-efficient batch-recirculated photoreactor packed with immobilized TiO2-P25 nanoparticles onto glass beads for photocatalytic degradation of phenazopyridine as a pharmaceutical contaminant: artificial neural network modeling*. Water Sci Technol, 2016. **73**(11): p. 2804–14.

[98] Akkari, M., et al., *ZnO/sepiolite heterostructured materials for solar photocatalytic degradation of pharmaceuticals in wastewater*. Appl Clay Sci, 2018. **156**: p. 104–109.

[99] Kumar, A., et al., *Magnetically recoverable ZrO2/Fe3O4/Chitosan nanomaterials for enhanced sunlight driven photoreduction of carcinogenic Cr (VI) and dechlorination & mineralization of 4-chlorophenol from simulated waste water*. RSC Adv., 2016. **6**.

[100] Sheydaei, M., et al., *Preparation of nano N-TiO2/graphene oxide/titan grid sheets for visible light assisted photocatalytic ozonation of cefixime*. Chem Eng J, 2018. **353**: p. 138–146.

[101] Deng, S.T., et al., *Enhanced degradation of ciprofloxacin by cerium oxide/graphene oxide nanocomposite under ultrasound irradiation*. Fuller Nanotub Carbon Nanostructures, 2021: p. 1–5.

[102] Kaur, A. and S. Kansal, *Bi2WO6 nanocuboids: An efficient visible light active photocatalyst for the degradation of levofloxacin drug in aqueous phase*. Chem Eng J, 2016. **302**.

[103] Hassani, A., et al., *Degradation of mixture of three pharmaceuticals by photocatalytic ozonation in the presence of TiO2/montmorillonite nanocomposite: Simultaneous determination and intermediates identification*. J Environ Chem Eng, 2017. **5**(2): p. 1964–1976.

[104] Hu, X.-Y., et al., *Graphene/TiO2/ZSM-5 composites synthesized by mixture design were used for photocatalytic degradation of oxytetracycline under visible light: Mechanism and biotoxicity*. Appl Surf Sci, 2016. **362**: p. 329–334.

[105] Li, J., et al., *Visible light driven photocatalytic decomposition of penicillin G by Ti 3+ self-doped TiO 2 nano-catalyst through response surface methodology*. J Taiwan Inst Chem Eng, 2018. **87**.

[106] Liu, X., et al., *Performance and mechanism into TiO2/Zeolite composites for sulfadiazine adsorption and photodegradation*. Chem Eng J, 2018. **350**: p. 131–147.

[107] Trawiński, J., R. Skibiński, and P. Szymański, Investigation of the photolysis and TiO(2), SrTiO(3), H(2)O(2)-mediated photocatalysis of an antipsychotic drug loxapine – Evaluation of kinetics, identification of photoproducts, and in silico estimation of properties. Chemosphere, 2018. **204**: p. 1–10.

Awais Ahmad, Hina Sharif, Rafael Luque, Mabkhoot Alsaiari, Farid A. Harraz

7 Water splitting using bionanomaterials

Abstract: In order to address the challenging need of energy, climate change and environmental protection, the production of photocatalytic hydrogen is, relatively, the best choice due to its abundance. Various nanomaterials and bionanomaterials, such as conjugated polymers (CPs), have been recently been found to be promising alternatives for photocatalysts due to their unique properties, including economical, high stability, tunable optoelectronic properties and efficient physicochemical properties. This chapter summarizes the introduction of photocatalytic water splitting, the use of CPs as photocatalysts and their scope for advancement.

Keywords: Biomaterials, water splitting, CPs

7.1 Introduction

Hydrogen utilization is considered to be a sustainable and clean energy source in the coming era and its importance arises due to the generation of high energy upon combustion [1]. This source of energy is usually ecofriendly and free of any harmful byproducts. Currently, the production of hydrogen involves the production of carbon dioxide as a byproduct, thereby, alternative methods involving a limited production of any greenhouse gases is of research interest. The generation of photocatalytic hydrogen using solar energy on the surface of a photocatalyst offers one of the promising alternative methods for resolving the current energy crisis, without any harmful impact on the environment [2–4]. The Sun is an energy warehouse that irradiates a huge source of solar energy constantly towards the Earth and this source of energy is considered to be enough to meet the current expanding demand of humans [5]. Solar energy is considered as an inexpensive and copious source of energy but its utilization is still inefficient due to its intermittent and diffuse nature as well as its low density; thus, solar energy is often converted into chemical energy, which is an appealing solution to meet the current needs of energy worldwide [6]. Using solar energy, semiconducting nanomaterials are effectively used for photocatalysis and in photovoltaics. Being photovoltaic, these nanomaterials generate photocurrent due to charge migration; while being photocatalysts, these nanomaterials facilitate chemical conversion through the process of oxidation and reduction [7, 8].

Production of clean hydrogen fuel through photocatalytic water splitting techniques is one of the uses of solar energy, which offers promising results to overcome the global energy crises. This source of energy is clean, ecofriendly and renewable [9, 10]. The photocatalytic technology also helps to reduce carbon dioxide, thereby

https://doi.org/10.1515/9783110768749-007

simultaneously benefiting the environment [11, 12]. Photon energy is used for the process of photoexcitation in photocatalytic techniques, which often depends on the photocatalyst's optical gap. Visible light comprises about 53% of the solar energy; therefore, crucial photocatalysts with an optical gap below 3.0 eV is needed, which can enhance the absorption of sunlight in the visible region [5]. The use of photocatalysts for the generation of photocatalytic hydrogen was first introduced in 1972 and thereafter, several semiconductors have been explored and used for the efficient production of hydrogen [13]. To identify materials that have the optimal bandgap, which can promote efficient charge separation, stability against photocorrosion and allow for use in multiple applications is of research interest in order to gain excellence in hydrogen evolution technology. Moreover, the morphological dimensions and the surface area of photocatalysts are considered as key aspects [1, 14].

Over the last few decades, researchers and scientists have intensively devoted their research efforts in perusing effective photocatalysts for use in real world situations. Bionanomaterials, particularly p-conjugated polymers, have been proved recently as promising candidates for photocatalytic applications. These photocatalysts are environment friendly, abundantly present, semiconductor in nature, have tunable energy levels for both reduction and oxidation reactions, facilely prepared, show extraordinary chemical resistance to photobleaching and their molecular structures can be modified for better visible light utilization [5]. The diverse modification capabilities of the molecular structures of conjugated polymers (CPs) can generate many exciting photocatalytic compounds [15]. For achieving the maximum efficiency of CPs, ideal properties of CPs are required, including excellent absorption, efficient charge separation, properly matched energy levels and transport of electrons. To obtain these desirable properties, several building blocks have been developed [16]. Polyaniline, polyacetylene and polypyrrole, and their derivatives, are the significant CPs that possess potential intrinsic conductivity, while polyflorenes, polyphenylenes, polythiophenes and polyarylenevinylenes are significant due to their photoluminescence and electro-optical properties. In contrast to all these, the nanoparticles of CPs show promising efficacy and photocatalytic activity [17–19].

7.1.1 Splitting of water by photocatalysis

In photocatalytic water splitting, hydrogen is usually generated via the irradiation process on the surface of a photocatalyst using solar energy. These photocatalysts represent the semiconductor materials that cause the water splitting without being used or transformed throughout the process. Due to the irradiation on the surface of photocatalysts, electron–hole pairs are created that are usually quite unstable and have a potential of easy recombination [1]. The activity of photocatalysts for the process of water splitting is usually dependent on their physicochemical properties, the reaction conditions and the nature of the co-catalysts [20, 21]. In the last few years, several compounds have been reported as photocatalysts that are capable of producing hydrogen and oxygen using visible light [1, 20, 22]. Figure 7.1 gives the list of the desirable features of photocatalysts to be used for effective hydrogen generation.

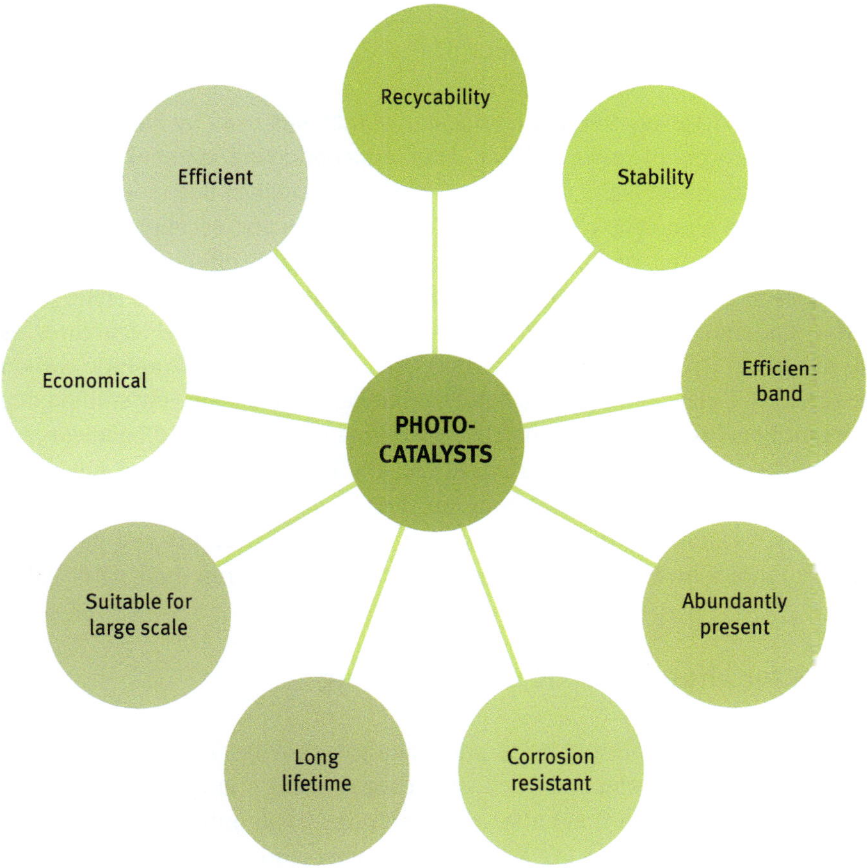

Figure 7.1: Ideal features of photocatalysts for hydrogen evolution.

The process of splitting of water by photocatalysis is often referred to as synthetic photosynthesis as this process is much similar to the process of photosynthesis in plants using sunlight [23]. The multi-electron and endothermic nature of photocatalytic splitting of water makes it a cycle. The conversion of water into hydrogen and oxygen needs an energy source to fulfill the Gibbs free energy that is needed to rearrange the valence electrons in the water molecules [24]. Photocatalytic hydrogen evolution offers several advantages over traditional sources of energy, such as (a) excellent conversion of solar energy, (b) effectively produced from water, (c) abundantly present sources, (d) can be stored in various storage conditions (both in liquid or gas form), (e) easily transferable over long areas, (f) can be converted into several forms of fuels and (g) completely environment friendly [25]. Photocatalysis is one of the methods that can be used to generate hydrogen using visible light. The reaction mechanism involved includes both reduction and oxidation processes;

$$\text{Photoreduction: } 2H_2O + 2e^- \rightarrow H_2 + 2OH^-$$
$$\text{Photo-oxidation: } 2H_2O \rightarrow O_2 + 4H^+ + 4e^-$$

Titanium oxide (TiO_2) was the first photocatalyst that was used by Fugishima and Honda for the process of water splitting. This semiconducting material has gained significant attention and is the most investigated photocatalyst due to its extraordinary stability and photolytic charge separation properties and excellent absorption [23]. Metal loading, such as Pt, Pd, Au, Ni and Ag onto TiO_2, offers surface modification for decreasing the recombination of photogenerated electron–hole pairs [26–28]. Graphene, due to its electrochemical properties and novel structures, is also used effectively for photocatalytic hydrogen evolution [29]. Graphitic carbon nitride (g-C_3N_4) is also another photocatalyst that has been extensively studied due to its potent benefits such as nontoxicity, metal-free semiconducting behavior and narrow bandgaps. Moreover, it is economical, abundantly present and has resistance to photocorrosion due to the strong covalent bonds between the nitride atoms and carbon [30–32]. Unfortunately, the use of this photocatalyst is less due to its low specific surface area and its high recombination with electron–hole pairs [23].

7.1.2 CPs for photocatalytic water splitting

The technique of water splitting using photocatalysis provides alternate and effective pathways for the production of clean hydrogen using sunlight and water [5]. The ultimate hydrogen obtained after water splitting is an optimal fuel having a gravimetric energy density of 142 MJ/kg (three times more than gasoline). Moreover, the combustion of H_2 results in no harmful impact to the environment and is completely ecofriendly as compared to gasoline. Fujishima and Honda were the pioneers in reporting water splitting on TiO_2 photoelectrodes, and thereafter several semiconductors have been studied and investigated for hydrogen production [33–35]. It is anticipated that solar hydrogen produced in this way can play a significant role in the prospective sustainable energy, as it is transportable and storable, and hence can be converted into electricity using fuel cells. Moreover, hydrogen produced via water splitting can be used as a feedstock in chemical industries and also for the recycling of carbon dioxide [10].

As found to be promising photocatalysts, CPs have been extensively studied and utilized for the production of hydrogen in the early 1980s, but their use remained restricted until the demonstration of three-dimensional (3D) poly(azomethine) networks in 2010 [36]. After this discovery, several exciting and potent organic CPs have been produced in order to improve the efficiency of H_2 production. Using the optimization techniques, several photophysical properties of CPs have been optimized, including bandgap alignment, harvesting of light in the visible region and photocharge generation [5]. CPs offer a number of promising benefits for the production of photocatalytic hydrogen due to its low-cost fabrication, efficient flexibility, versatile functionalization and speedy scale-up of production [16]. The pi-electrons delocalization of CPs make them preferable for light harvesting and also for efficient proton reduction. Moreover, CPs are generally nontoxic and are resistant to photcorrosion. Several classes of CPs have been developed and optimized, such as conjugated porous polymers (CPPs), covalent organic frameworks (COFs), covalent triazine frameworks (CTFs) and polymer dots, which have been proved to be efficient photocatalysts [37].

7.1.3 Linear conjugated polymers

Reports of water splitting generated via linear organic CPs as photocatalysts can be observed in the twentieth century. Shibata et al. [38] used poly(*p*-phenylene) (PPP), prepared through the polycondensation of 1,4-dibromobenzene, for water splitting (Figure 7.1(a)). This linear conjugated polymer has a bandgap of 2.9 eV and its hydrogen evolution rate (HER) was found to be 207.5 mmL/h g. But this polymer was inactive under visible light (4,400 nm). After this work, several analogues of PPP were reported, but still low activities were achieved [39, 40]. In the early twenty-first century, extended planarized units were incorporated into the main structure of PPP for gaining evolution in H_2 production, which was then expected to promote the reactions of proton reduction (Figure 7.1(b)) [41]. The HER obtained from such new CPs were found to be about 128 mmol/h g, which is 3 times more than that obtained from PPP. Particularly the incorporation of SO_2 subunit in the position of X, as shown in Figure 7.1, showed an astonishing HER 1,492 μmol/h g, which was about 46 times higher than that of PPP [41]. After the results were obtained from these CPs, ethnyl group was incorporated into the main chain, which further reduced the bandgap. This polymer showed an impressive HER value of 6,023 mmol/h g [42]. Later on, the polymers d and e (Fig. 3) were produced, which showed HER values of 4,270 and 3,187 mmol/h g, respectively [5].

In order to expand the conjugation of polymers, large planar units were also incorporated into the main chain, such as polymer (f) in Figure 7.1, whose bandgap was found to be 2.07 eV and HER was 5,040 mmol/h g [43]. The production of soluble CPs offers a better processing of polymers for the photocatalytic production of H_2 as compared to the insoluble bulk polymers. Moreover, the production of polymer dots offers smaller size, thereby shortening the distance between the photogenerated charges and the photocatalyst particles, and in this way, the probability of recombination of holes and electrons ultimately reduces and improves the photocatalytic efficiency [7, 44, 45].

7.1.4 Conjugated porous polymers (CPPs)

The cross-linking of CPs offering 3D porous features has gained significant importance due to the large electron delocalized systems, tunable optoelectronic properties and excellent stability. These features have invigorated the wide applications of CPs in energy storage, gas absorption heterogeneous catalysis and chemical sensing [46, 47]. Along with these benefits, CPPs have also gained significant research interest for the photocatalytic H_2 evolution in this era. The porous structures of CPPs offer an elevated interaction of water molecules with polymers and hence promoting the charge transfer reactions, which ultimately magnifies the photocatalytic reactions [48]. The copolymerization of aromatic monomers has been found to boost the catalytic activity of CPPs via improving bandgap, charge mobility and porosity [5]. Recently, conjugated microporous polymers have been produced that exhibit high surface areas and bandgaps, and the HER values have been found to be astonishing [5]. The approach of the donor-acceptor combination is another prevailing method that can be used in polymer solar cells to improve the efficiency of power conversion techniques. The interaction of intramolecular charge transfer can be induced effectively with alternating electron-deficient acceptor (A) and electron-rich donor (D) in the main chain, which effectively reduces the bandgap and modulates the molecular energy. This approach has significant interest in designing and manufacturing several CPs for the technique of photocatalytic H_2 evolution [49–52]. Liu et al. [53] reported CPPs that were produced using a combination of chromophores with bipyridyl and biphenyl. The introduction of chromophore centers in the CPPs allowed the lowering of bandgaps to a significant level, leading to an overall H_2 evolution [53].

The substitution of fluorine atom (F) in the CPs backbone is also an alternative and effective approach to enhance the photoelectronic properties of CPs. The F introduction reduces the steric hindrance but profoundly influences the molecular orbits of CPs [54]. Xiang et al. [55] reported the effect of fluorine substitution, both in linear CPs and CPPs, and observed that there was a weak improvement in the optical gaps, but the efficiency of charge transfer was improved significantly. In this way, the HER value was enhanced 5-fold than the nonfluorinated counterpart. Generally, the extension of para conjugation imposes several impacts on the porous properties, photophysical nature and the morphologies of the resulting polymers, which ultimately result in a synergistic impact on the photocatalytic nature of CPs [5].

7.2 Covalent organic frameworks (COFs)

COFs represent an emerging and exciting class of 2D and 3D polymers, which are formed by the linkage of organic building units into the extended structures that are held together by strong covalent bonds [56, 57]. The efficient catalytic activity of COFs is usually governed due to their high crystallinity, which favors effective charge separation and transport. In addition, the excellent stability, tunable physicochemical properties and well-defined pores of COFs make them an attractive choice for photocatalytic H_2 production [58]. The modification of substituents in the COF also offers good photocatalytic performances [5].

7.3 Covalent triazine frameworks (CTFs)

CTFs are another class of organic framework materials that offer high nitrogen content and unique porosity, and both of these features are tunable by the suitable selection of organic linkers. In the recent era, CTFs have found extensive application in capturing carbon dioxide from the environment, energy storage and photocatalysis [59–61]. CTFs are synthesized using triazine-based networks through the self-condensation of nitriles, which result in highly porous materials. 1,4-Dicyanobenzene was the first triazine network that possessed a high degree of crystallinity [62]. Bi et al. [63] prepared CTF through the trimerization reaction of terephthalonitrile, which exhibited a bandgap of 2.94 eV and a HER of 200 mmol/h g. Next, Kuecken et al. [64] reported that CTF having less HER values can be converted into effective photocatalysts by mixing the CTF in chloroform and zinc chloride salt for about ten minutes, which showed an optical gap of 2.42 eV and a HER value of 1,072 mmol/h g. Another disadvantage of CTFs is the rapid recombination of CTFs holes and the photogenerated electrons, which reduces the high efficiency. But several investigations have been done to optimize CTFs in order to obtain an efficient charge carrier density and a high photocurrent, which ultimately improve the charge separation efficiency [65].

7.4 Conclusion

Owing to the diversified molecular structures and effortlessly tunable physicochemical properties of CPs, these exciting bionanomaterials offer unlimited opportunities and challenges in achieving a high efficiency of water splitting technology, which is believed to be an alternative and a huge source for meeting the current expanding need of energy globally. We expect that this chapter will prove to be a good reference in future research in this emerging field. With the advancements in the material science and polymer chemistry, it is expected that innovations and optimization in CPs can play a significant role in improving the efficiency of the photocatalytic technology.

References

[1] Ganguly, P., et al., *2D Nanomaterials for Photocatalytic Hydrogen Production*. ACS Energy Letters, 2019. **4**(7): p. 1687–1709.

[2] Chiesa, P., et al., *Co-production of hydrogen, electricity and CO2 from coal with commercially ready technology. Part A: Performance and emissions*. International Journal of Hydrogen Energy, 2005. **30**(7): p. 747–767.

[3] Rostrup-Nielsen, J.R. and T. Rostrup-Nielsen, *Large-Scale Hydrogen Production*. CATTECH, 2002. **6**(4): p. 150–159.

[4] Chalk, S.G. and J.F. Miller, *Key challenges and recent progress in batteries, fuel cells, and hydrogen storage for clean energy systems*. Journal of Power Sources, 2006. **159**(1): p. 73–80.

[5] Dai, C. and B. Liu, *Conjugated polymers for visible-light-driven photocatalysis*. Energy and Environmental Science, 2019. **13**.

[6] Carrillo, A.J., et al., *Solar energy on demand: a review on high temperature thermochemical heat storage systems and materials*. Chemical Reviews, 2019. **119**(7): p. 4777–4816.

[7] Kundu, S. and A. Patra, *Nanoscale Strategies for Light Harvesting*. Chemical Reviews, 2017. **117**(2): p. 712–757.

[8] Ahmad, A., et al., *A novel study on synthesis of egg shell based activated carbon for degradation of methylene blue via photocatalysis*. Arabian Journal of Chemistry, 2020. **13**(12): p. 8717–8722.

[9] Chen, X., et al., *Semiconductor-based photocatalytic hydrogen generation*. Chemical Reviews, 2010. **110**(11): p. 6503–6570.

[10] Hisatomi, T., J. Kubota, and K. Domen, *ChemInform Abstract: recent advances in semiconductors for photocatalytic and photoelectrochemical water splitting*. Chemical Society Reviews, 2014. **43**.

[11] Habisreutinger, S.N., L. Schmidt-Mende, and J.K. Stolarczyk, *Photocatalytic reduction of CO2 on TiO2 and other semiconductors*. Angewandte Chemie International Edition England, 2013. **52**(29): p. 7372–408.

[12] Liu, X., S. Inagaki, and J. Gong, *Heterogeneous Molecular Systems for Photocatalytic CO(2) Reduction with Water Oxidation*. Angewandte Chemie International Edition England, 2016. **55**(48): p. 14924–14950.

[13] Zhou, X., et al., *Noble-metal-free photocatalytic hydrogen evolution activity: the impact of ball milling anatase nanopowders with TiH2*. Advanced Materials, 2017. **29**(5): p. 1604747.

[14] Ganguly, P., et al., *Antimicrobial activity of photocatalysts: Fundamentals, mechanisms, kinetics and recent advances*. Applied Catalysis B: Environmental, 2018. **225**: p. 51–75.

[15] Krivtsov, I., et al., *Water-soluble polymeric carbon nitride colloidal nanoparticles for highly selective quasi-homogeneous photocatalysis*. Angewandte Chemie International Edition, 2019. **59**.

[16] Zhang, G., Z.-A. Lan, and X. Wang, *Conjugated Polymers: Catalysts for Photocatalytic Hydrogen Evolution*. Angewandte Chemie International Edition 2016. **55**(51): p. 15712–15727.

[17] Günes, S., H. Neugebauer, and N.S. Sariciftci, *Conjugated polymer-based organic solar cells.* Chemical Reviews, 2007. **107**(4): p. 1324–1338.

[18] Mishra, A., C.-Q. Ma, and P. Bäuerle, *Functional oligothiophenes: molecular design for multidimensional nanoarchitectures and their applications.* Chemical Reviews, 2009. **109**(3): p. 1141–1276.

[19] Pecher, J. and S. Mecking, *Nanoparticles of conjugated polymers.* Chemical Reviews, 2010. **110**(10): p. 6260–6279.

[20] Maeda, K. and K. Domen, *Photocatalytic Water Splitting: Recent Progress and Future Challenges.* Journal of Physical Chemistry Letters, 2010. **1**(18): p. 2655–2661.

[21] Aravind, M., et al., *Critical green routing synthesis of silver NPs using jasmine flower extract for biological activities and photocatalytical degradation of methylene blue.* Journal of Environmental Chemical Engineering, 2021. **9**(1): p. 104877.

[22] Wang, Q. and K. Domen, *Particulate photocatalysts for light-driven water splitting: mechanisms, challenges, and design strategies.* Chemical Reviews, 2020. **120**(2): p. 919–985.

[23] Fajrina, N. and M. Tahir, *A critical review in strategies to improve photocatalytic water splitting towards hydrogen production.* International Journal of Hydrogen Energy. 2019. **44**(2): p. 540–577.

[24] Maeda, K., *Photocatalytic water splitting using semiconductor particles: History and recent developments.* Journal of Photochemistry and Photobiology, 2011. **12**(4): p. 237–258.

[25] Acar, C., I. Dincer, and G.F. Naterer, *Review of photocatalytic water-splitting methods for sustainable hydrogen production.* International Journal of Energy Research, 2016. **40**(11): p. 1449–1473.

[26] Low, J., B. Cheng, and J. Yu, *Surface modification and enhanced photocatalytic CO2 reduction performance of TiO2: a review.* Applied Surface Science, 2017. **392**: p. 658–686.

[27] Kashif, M., et al., *An experimental and DFT study on novel dyes incorporated with natural dyes on titanium dioxide (TiO2) towards solar cell application.* Applied Physics A, 2020. **126**(9): p. 716.

[28] Saleem, M., et al., *Experimental and theoretical study of highly porous lignocellulose assisted metal oxide photoelectrodes for dye-sensitized solar cells.* Arabian Journal of Chemistry, 2021. **14**(2): p. 102937.

[29] Kashif, M., et al., *Effect of potassium permanganate on morphological, structural and electro-optical properties of graphene oxide thin films.* Arabian Journal of Chemistry, 2021. **14**(2): p. 102953.

[30] Cao, S., et al., *Polymeric photocatalysts based on graphitic carbon nitride.* Advanced Materials, 2015. **27**(13): p. 2150–2176.

[31] Liu, J., et al., *Simple pyrolysis of urea into graphitic carbon nitride with recyclable adsorption and photocatalytic activity.* Journal of Materials Chemistry, 2011. **21**: p. 14398.

[32] Bibi, S., et al., *Photocatalytic degradation of malachite green and methylene blue over reduced graphene oxide (rGO) based metal oxides (rGO-Fe3O4/TiO2) nanocomposite under UV-visible light irradiation.* Journal of Environmental Chemistry England, 2021. **9**: p. 105580.

[33] Karayilan, M., et al., *Catalytic Metallopolymers from [2Fe-2S] Clusters: Artificial Metalloenzymes for Hydrogen Production.* Angewandte Chemie International Edition England, 2019. **58**(23): p. 7537–7550.

[34] Ma, Y., et al., *Titanium dioxide-based nanomaterials for photocatalytic fuel generations.* Chem Rev, 2014. **114**(19): p. 9987–10043.

[35] Wu, L.Z., et al., *Enhancement of the efficiency of photocatalytic reduction of protons to hydrogen via molecular assembly.* Accounts of Chemical Research, 2014. **47**(7): p. 2177–85.

[36] Schwab, M., et al., *Photocatalytic hydrogen evolution through fully conjugated poly (azomethine) networks*. ChemComm, 2010. **46**: p. 8932–4.

[37] Vyas, V.S., V.W. Lau, and B.V. Lotsch, *Soft photocatalysis: organic polymers for solar fuel production*. Chemistry of Materials, 2016. **28**(15): p. 5191–5204.

[38] Shibata, T., et al., *Novel visible-light-driven photocatalyst. Poly(p-phenylene)-catalyzed photoreductions of water, carbonyl compounds, and olefins*. Journal of Physical Chemistry, 1990. **94**(5): p. 2068–2076.

[39] Pac, C., et al., *A novel type of photosubstitution of [fac-Re(bpy)(CO)3Br](bpy = 2,2′-bipyridine) with P(OR)3 in the presence of triethylamine: efficient formation of [Re(bpy)(CO)2{P(OR)3}2] +Br–*. Journal of Chemical Society, 1991(11): p. 787–788.

[40] Matsuoka, S., et al., *Photocatalysis of oligo(p-phenylenes): photoreductive production of hydrogen and ethanol in aqueous triethylamine*. Journal of Physical Chemistry, 1991. **95**(15): p. 5802–5808.

[41] Sprick, R.S., et al., *Visible-light-driven hydrogen evolution using planarized conjugated polymer photocatalysts*. Angewandte Chemie International Edition, 2016. **55**(5): p. 1792–1796.

[42] Xiang, Y.G., et al., *Enhanced visible light photocatalytic activity of TiO_2 assisted by organic semiconductor: a structure optimization strategy of conjugated polymer*. Journal of Materials Chemistry, 2017. **6**.

[43] Dai, C., et al., *Dibenzothiophene-S,S-dioxide-based conjugated polymers: highly efficient photocatalyts for hydrogen production from water under visible light*. Small, 2018. **14**(34): p. e1801839.

[44] Pecher, J. and S. Mecking, *Nanoparticles of conjugated polymers*. Chemical Reviews, 2010. **110**(10): p. 6260–6279.

[45] Feng, L., et al., *ChemInform Abstract: conjugated polymer nanoparticles: preparation, properties, functionalization and biological application*. Chemical Society Reviews, 2013. **42**.

[46] Dawson, R., A.I. Cooper, and D.J. Adams, *Nanoporous organic polymer networks*. Progress in Polymer Science, 2012. **37**(4): p. 530–563.

[47] Zhu, J., et al., *Two-dimensional porous polymers: from sandwich-like structure to layered skeleton*. Accounts of Chemical Research, 2018. **51**(12): p. 3191–3202.

[48] Vilela, F., K. Zhang, and M. Antonietti, *Conjugated porous polymers for energy applications*. Energy and Environmental Science, 2012. **5**: p. 7819–7832.

[49] Li, Y., *Molecular design of photovoltaic materials for polymer solar cells: toward suitable electronic energy levels and broad absorption*. Accounts of Chemical Research, 2012. **45**(5): p. 723–733.

[50] Zhou, H., L. Yang, and W. You, *Rational design of high performance conjugated polymers for organic solar cells*. Macromolecules, 2012. **45**(2): p. 607–632.

[51] Yao, H., et al., *Molecular design of benzodithiophene-based organic photovoltaic materials*. Chemical Reviews, 2016. **116**(12): p. 7397–7457.

[52] Gao, C., et al., *Rational design on D–A conjugated P(BDT–DTBT) polymers for polymer solar cells*. Polymer Chemistry, 2014. **5**.

[53] Liu, C., et al., *Low bandgap semiconducting polymers for polymeric photovoltaics*. Chemical Society Reviews, 2015. **45**.

[54] Yue, J., et al., *Effects of pyridyl group orientations on the optoelectronic properties of regio-isomeric diketopyrrolopyrrole based π-conjugated polymers*. Journal of Materials Chemistry, 2016. **4**.

[55] Xiang, Y., et al., *Conjugated polymers with sequential fluorination for enhanced photocatalytic H_2 evolution via proton-coupled electron transfer*. ACS Energy Letters, 2018. **3**(10): p. 2544–2549.

[56] Segura, J., M.J. Mancheño, and F. Zamora, *Covalent organic frameworks based on Schiff-base chemistry: Synthesis, properties and potential applications*. Chemical Society Reviews, 2016. **45**.
[57] Mastalerz, M., *The next generation of shape-persistent zeolite analogues: covalent organic frameworks*. Angewandte Chemie International Edition England, 2008. **47**(3): p. 445–7.
[58] Beuerle, F. and B. Gole, *Covalent organic frameworks and cage compounds: design and applications of polymeric and discrete organic scaffolds*. Angewandte Chemie International Edition England, 2018. **57**(18): p. 4850–4878.
[59] Subodh, et al., *A new triazine-cored covalent organic polymer for catalytic applications*. Applied Catalysis A: General, 2020. **593**: p. 117411.
[60] Thomas, A., *Functional materials: from hard to soft porous frameworks*. Angewandte Chemie International Edition, 2010. **49**(45): p. 8328–8344.
[61] Sakaushi, K. and M. Antonietti, *Carbon- and nitrogen-based organic frameworks*. Accounts of Chemical Research, 2015. **48**(6): p. 1591–600.
[62] Kuhn, P., M. Antonietti, and A. Thomas, *Porous, covalent triazine-based frameworks prepared by ionothermal synthesis*. Angewandte Chemie International Edition England, 2008. **47**(18): p. 3450–3.
[63] Bi, J., et al., *Covalent triazine-based frameworks as visible light photocatalysts for the splitting of water*. Macromolecular Rapid Communication, 2015. **36**(20): p. 1799–805.
[64] Kuecken, S., et al., *Fast tuning of covalent triazine frameworks for photocatalytic hydrogen evolution*. ChemComm 2017.
[65] Zhou, Q., et al., *A clustering-triggered emission strategy for tunable multicolor persistent phosphorescence*. Chemical Science, 2020. **11**(11): p. 2926–2933.

Awais Ahmad, Hina Sharif, Rafael Luque, Arsh-e- Noor

8 Biomaterials: a good source for H_2 production via photocatalysis

Abstract: Biomaterials offer a sustainable and renewable source of photocatalysts. It has allowed for the development of extraordinary applications in sustaining the ecosystem as well as for the well-being of humans. Due to overpopulation, increasing energy demand and issues of global warming throughout the world, renewable and clean alternative energy sources have gained significant attention in the field of science and research. Solar energy is an inexhaustible energy source that can be used for hydrogen production via water splitting. For this purpose, photocatalysts have been developed, and biomaterials can also be used as photocatalysts. For the efficient hydrogen production on a large scale, the available photocatalysts are being modified using co-catalysts and/or doping techniques, which can enhance the photocatalytic properties of the photocatalysts. In this chapter, we reviewed some photocatalysts and the various approaches that have been made to enhance their photocatalytic properties, along with the future prospects.

Keywords: Biomaterials, H_2 production, photocatalysis, energy applications

8.1 Introduction

Biomaterials showing multi functions are becoming surprisingly significant in the current technology innovations, showing extraordinary outcomes. These miniature and futuristic materials of nature have always amazed the scientific community for sustaining the ecosystem and well-being of humans in accordance with the rules of green chemistry [1]. The current world is facing challenging problems of environmental pollution, infectious diseases and climate change. These environmental conditions have led to the exhaustion of non-renewable energy sources and the environmental crises [2]. The lack of non-renewable energy sources or fossil fuels has enhanced scientific interest in the use of biomaterials and greener technology, for alleviating pollution and for production of clean energy. Hydrogen, being a benign combustion product and having extraordinary enthalpy value, is regarded as a clean and prospective source of energy as compared to the traditional fossil fuels. The process of water splitting via photocatalysis is usually economical and environmental friendly. The process of photocatalysis has recently gained interest in water treatment and disinfection due to its high efficiency [1, 3–5]. Photocatalysis is the efficient alternative for renewable production of hydrogen, but it requires new catalysts that give high efficiency at a large scale [6]. Popular metallic photocatalysts

https://doi.org/10.1515/9783110768749-008

have offered sustainable and better performances due to the gradual shift of inorganic nano-materials toward organic- and polymer-based photocatalysts [1, 7].

The harmful environmental intoxication and energy crises, currently, due to the rapid exhaustion of natural gas, oil and coal has emphasized on exploiting the renewable, clean and eco-friendly energy sources [4]. Due to the increasing demand of energy and the environmental problems, the production of hydrogen from water spitting via photocatalysis is a research interest as an artificial photosynthetic process (Figure 8.1) [8, 9]. The solar radiations that reach the Earth's surface consist of near-infrared (NIR) light (52%–55%), visible light (42%–43%) and UV light (3–5%). Keeping in view the spectrum of these solar radiations, the development of stable and efficient photocatalysts is becoming a high priority [10, 11]. Regarding the environment protection interests, a photocatalyst should be durable, free from secondary pollutants in scale-up applications and totally eco-friendly. To gain these concerns, metal-free photocatalysts are often preferred that show no metal ion leakage [12]. Therefore, eco-friendly and biocompatible high performance metal-free photocatalysts are preferred to gain maximum utilization of the full spectrum energy. Hence, the modification of semiconductor photocatalysts, such as Z-scheme and metal ion doping, offers an efficient and desirable usage of energy [13, 14].

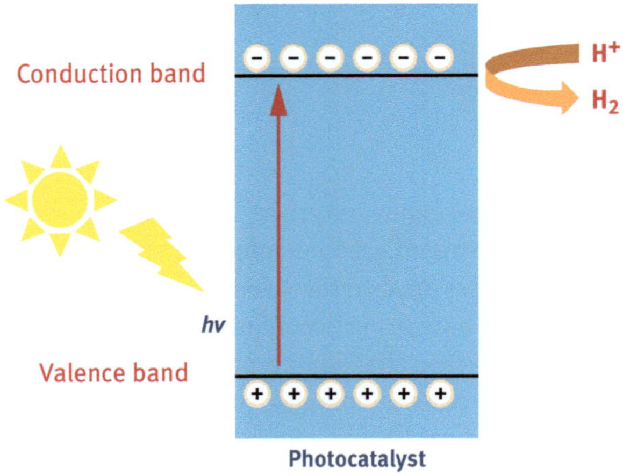

Figure 8.1: H_2 production through photocatalysis.

8.1.1 Photocatalytic hydrogen production

There are several primary energy sources from which hydrogen can be produced, such as natural gas, coal, biomass, waste materials, solar light, nuclear power, geothermal power and hydropower. Commercial hydrogen is usually produced from coal using the gastification technique and also from steam, reforming methane, but both these methods also involve the emission of carbon dioxide [15, 16]. Hydrogen is an effective alternative source of energy that can be stored in liquid, gaseous and/or metal hydride form, and can also be distributed easily through pipelines or tankers over large distances [15]. To achieve an efficient and affordable production of hydrogen, water and solar light are the effective alternative sources. Hence, photocatalytic water splitting for hydrogen production offers a clean source of hydrogen and can be used in small as well as in large-scale generators for hydrogen production [17]. Honda and Fujishima were the pioneers who introduced photocatalytic water splitting in 1972 using Pt as the cathode and TiO_2 as the photoanode, and after this investigation, several studies and researches have been conducted on water splitting [18].

8.1.2 Biomaterials as photocatalysts for hydrogen production

The concept of photocatalysis has recently gained huge emphasis in undergraduate chemistry courses. Many researches and investigations have been reported regarding the degradation of pollutants using heterogeneous photocatalysts [19, 20]. The availability of advanced and novel nanomaterials is the major driving force for processes from hydrogen generation to energy propulsion [21]. Fujishima was the pioneer who introduced the use of TiO_2 for hydrogen production using the water splitting technique. Later, a number of photocatalytic materials have been discovered and synthesized for the efficient production of hydrogen and for pollutant decomposition [6, 22, 23].

8.2 TiO_2

Several nanomaterials can be used for photocatalytic hydrogen production, such as CdS, TiO_2 and SiC, as all of these photocatalysts show excellent photocatalytic properties [18]. TiO_2 is one of these photocatalysts that has been tremendously studied due to its excellent band-edge position, resistance to photo-corrosion, extraordinary stability and optoelectronic properties and for being economical. Various modifications are made to TiO_2 to lower its band gap, which enhances the adsorption of visible light [24, 25]. The addition of sacrificial reagents or electron donors (such as formic acid, oxalic acid and formaldehyde) in the TiO_2 is an effective measure that can increase the separation of electron–holes. However, to continue this reaction, there is a need for continuously adding electron donors, as they are used during the photocatalytic reactions [26, 27].

The alteration of the nanomaterial morphology, doping with metallic and/or nonmetallic elements (such as Pd, Pt, Rh, Ni Au and Ag) and the compounding of TiO_2

with other molecules are effective ways of optimizing the photocatalytic performance of TiO_2 [28–30]. Through these treatments, enhanced active sites, large surface areas and increased separation of photogenerated electron holes pairs can be achieved, which ultimately lead to an enhanced absorption intensity of the visible light. Hydrogenated TiO_2, therefore, has gained significant attention due to its enhanced photocatalytic activity [31]. The loading of TiO_2 with metals (Ag or Au) can also enhance the photocatalytic activity through surface plasmon resonance (SPR) effect. SPR refers to the overall motion of conducting electrons that is induced due to light irradiation, considerably enhancing the electric near field [32]. Loading of TiO_2 with Au results in the excitation of valence bond electrons to the conduction band through UV light irradiation, and which then are transferred to the Au particles on the surface of TiO_2. In this way, the SPR effect boosts up the energy intensity of electrons, ultimately increasing the photocatalytic activity [33]. Due to several limitations of TiO_2 as a photocatalyst for water splitting, TiO_2 is combined with two or more components to enhance its photocatalytic activity for commercialization, for example, NiO-$SrTiO_3$ and NiO-Ni are loaded with $Sr_2Nb_2O_7$ and $Sr_2Ta_2O_7$ [34].

8.3 Dimensional nanomaterials

Two-dimensional nanomaterials have been introduced as photocatalysts for the process of water splitting, which can reduce the recombination of electron–hole pairs and can also reduce the traveling distance of the charge carriers. This is accomplished due to the layered structure of 2D nanomaterials, which assist in the maximum light absorption at a low density [35, 36]. The 2D materials offer extraordinary optical as well as electronic properties, and their covalent bonds offer the generation of several hetero-structures and hetero-junctions. In addition to this, the 2D structure enhances the specific surface area, which increases the efficiency of water splitting. Graphene is one such 2D structure that shows exceptional electronic, optical and mechanical properties [37–39]. Similarly, two dimensional transition metal dichalcogenides (TMDs) have also been extensively studied in the last few years. They are made up of hexagonal metal atoms layers that are squeezed between two layers of chalcogen atoms [40]. In TMDs, the 2D structure is formed due to the covalent interactions of sheets while the 3D crystalline shape is formed due to van der Waals interactions between the adjacent layers. In this way, about 40 types of TMDs have been synthesized, based on the combination patterns of the chalcogen atoms and the transition metals [40, 41]. The utilization of 2D nanomaterials, along with potent co-catalysts, is an efficient and worthy option for the efficient production of hydrogen via the method of water splitting [42].

For the optimum synthesis of 2D nanomaterials, chemical vapor deposition (CVD) technique is preferred in which a solid structure is initially deposited on a substrate heated with vapor formation and then it is optimized done by tuning the substrate type, the substrate temperature and the pressure of the incoming gas [43]. This technique of optimizing 2D nanoarchitectures is quite reliable, uniform and efficient; nevertheless, this process is expensive as well due to the use of precious metal substrates [44]. Surface modulation through surface engineering is another useful technique to adjust the physicochemical properties of nanomaterials [45]. Yuan et al. [46] synthesized $Ti_{0.87}O_2$ and $Ca_2Nb_3O_{10}$ nanosheets for 2D oxide platelets growth, utilizing the wet chemical method and calcination, which enhanced the production of hydrogen impressively via water splitting. Instead of these methods, incorporating metal nanoparticles on the surface of semiconductors is an alternative and easier optimizing method, which can increase the basic properties of photocatalysts such as charge separation process, increasing the band gap absorption in the visible region and/or enhancing the photochemical process [42].

8.4 Graphitic carbon nitride (g-C_3N_4)

Graphitic carbon nitride (g-C_3N_4) has recently been found to be an eminent photocatalyst due to its excellent structure and band gap and it uses visible light for H_2 production [47, 48]. C_3N_4, in the form of melons, is considered as one of the traditional polymers but its allotropes are α-C_3N_4, β-C_3N_4, g-C_3N_4, cubic C_3N_4 and pseudocubic C_3N_4. g-C_3N_4 is the most stable allotrope under circumambient conditions [48, 49]. In spite of this, the composite photocatalyst made from g-C_3N_4 and NiS shows strong potential over g-C_3N_4 because NiS possesses the ability to impede the electrons and holes recombination and hence enhances H_2 production [50, 51]. Moreover, NiS is also economical as compared to the noble metal co-catalyst, Pt, and it adsorbs strong NIR region, ultimately extending the photo-response effectively [52]. In spite of these advantages, H_2 production is not satisfactory using the composite photocatalyst of g-C_3N_4 and NiS due to the random distribution and aggregation of NiS during the process of formation, which reduces the electrons and holes separation [53].

For an even distribution of NiS in g-C_3N_4, the coordination polymer of Ni^{2+}, with the thiol group containing an appropriate N-heterocycle molecule, is often an ideal option due to the conversion of Ni^{2+}-based coordination polymer in NiS@g-C_3N_4, through calcination [54]. In the coordination polymer, the N-heterocycle molecule and Ni^{2+} are connected in a high array, which insures the homogenize distribution of NiS, the nano-composite. Thus, NiS nanoparticles are produced due their limited movement by the proximate thiol groups having N-heterocycle molecules [55, 56]. Moreover, the presence of N-atom generates vacancy in the nanocomposite, which further facilitates the electron–hole separation and an increase in light adsorption in the NIR region of the photocatalyst [57]. Table 8.1 lists the photocatalysts that can be used for the production of hydrogen.

8.5 Microorganisms as photocatalytic water splitting

Photobiological water splitting is the method in which microorganisms like green algae and cyanobacteria are used for the production of hydrogen utilizing the solar light [58]. These microorganisms have the capability to absorb light using the pigments present in their photosystems. This absorbed light can then be effectively transferred to the centers of chlorophylls where several oxidants and reductants are formed via the process of charge separation [59, 60]. These generated oxidants and reductants can react in the photosystems and generate electron–hole pairs, leading to the process of water splitting. However, the overall production rate of hydrogen through this photobiological water splitting is low; thereby this method is commercially unviable, but this method of water splitting needs more research and investigation to ensure an environmentally safe source of energy [61].

Table 8.1: List of photocatalysts used for H_2 production.

Photocatalysts	Source of light	Rate of H_2 production (μmol/h)	Reference
Pt/TiO_2	400 W Hg lamp	568	[62]
ZrO_2	400 W Hg lamp	72–142	[63]
Pt/ZrO_2	400 W Hg lamp	120	[63]
$NiO/Sr_2Nb_2O_7$	400 W Hg lamp	110	[34]
$NiO/Sr_2Ta_2O_7$	400 W Hg lamp	1000	[34]
$BaTa_2O_6$	400 W Hg lamp	21	[64]
$BaTa_2O$	400 W Hg lamp	126	[64]
$NiO/BaTa_2O_6$	400 W Hg lamp	629	[64]
$Ni/Rb_4Nb_6O_{17}$	400 W Hg lamp	936	[65]
$Ni/K_4Nb_6O_{17}$	400 W Hg lamp	403	[65]
$Pt/BaTaO_2N$	300 W Xe lamp	6.6	[66]
$Pt/SrTiO_3$; Bi_2MoO_6	300 W Xe lamp	19	[67]
$Pt/SrTiO_3$: Rh WO_3	500 W halogen lamp	1.6	[68]
NiS/CdS	300 W Xe lamp	69.8	[69]
CdS/MoS_2	300 W Xe lamp	55	[70]
ZnS-g-C_3N_4	300 W Xe lamp	69.8	[71]

8.6 Current challenges and future prospects of water splitting

To overcome the increasing demand of energy and the saturation of energy resources, water splitting method is considered safe and efficient for the production of H_2, and this method can be enhanced by preventing the backward reactions of water splitting, separating the photo-generated charge separation and utilizing large incident energy fraction [26]. The reforming of photocatalytic process for the production of hydrogen at ambient conditions is the need of this era, as through this method, solar energy can be converted into chemical energy effectively. For this process, several chemical compounds (biomass and alcohols) have been extensively studied for photocatalytic reforming.

Several methods, such as doping and/or loading of metal or metal oxides on the surface of photocatalysts and synthesis of composite photocatalysts, can successfully improve the efficiency of a photocatalyst for the process of water splitting. These ways can tune the band gap and also prevent the recombination of electron–hole pairs. In short, the process of photocatalytic water splitting needs further research and investigations to explore the potential of semiconductors and/or biomaterials, which can then generate solar energy. More detailed and explanatory studies are required for the exploration of highly stable, eco-friendly, efficient and economical photocatalysts as well as modified photocatalysts to introduce clean hydrogen production on a large scale.

References

[1] Kumar, A., et al., *Bio-inspired and biomaterials-based hybrid photocatalysts for environmental detoxification: A review*. Chem. Eng. Sci., 2020. **382**: p. 122937.

[2] Cai, T., et al., *Recent advances in round-the-clock photocatalytic system: Mechanisms, characterization techniques and applications*. J. Photochem. Photobiol. C: Photochem. Rev., 2019. **39**: p. 58–75.

[3] Garcia-Segura, S. and E. Brillas, *Applied photoelectrocatalysis on the degradation of organic pollutants in wastewaters*. J. Photochem. Photobiol. C: Photochem. Rev., 2017. **31**: p. 1–35.

[4] Lu, L., et al., *Coordination Polymer Derived NiS@g-C3N4 Composite Photocatalyst for Sulfur Vacancy and Photothermal Effect Synergistic Enhanced H2 Production*. ACS Sustain. Chem. Eng., 2018. **6**(9): p. 11869–11876.

[5] Ahmad, A., et al., *A novel study on synthesis of egg shell based activated carbon for degradation of methylene blue via photocatalysis*. Arab. J. Chem., 2020. **13**(12): p. 8717–8722.

[6] Xiang, Q., et al., *Interfacial modification of titanium dioxide to enhance photocatalytic efficiency towards H2 production*. J. Colloid Interface Sci., 2019. **556**: p. 376–385.

[7] Aravind, M., et al., *Critical green routing synthesis of silver NPs using jasmine flower extract for biological activities and photocatalytical degradation of methylene blue*. J. Environ. Chem. Eng., 2021. **9**(1): p. 104877.

[8] Liu, Y., et al., *Ultrastable metal-free near-infrared-driven photocatalysts for H2 production based on protonated 2D g-C3N4 sensitized with Chlorin e6*. Appl. Catal. B. 2020. **260**: p. 118137.

[9] Yao, Y., et al., *Photocatalytic reforming for hydrogen evolution: a review*. Catalysts, 2020. **10**(3): p. 335.

[10] Zhu, M., et al., *Noble metal-free near-infrared-driven photocatalyst for hydrogen production based on 2D hybrid of black Phosphorus/WS2*. Appl. Catal. B. 2018. **221**: p. 645–651.

[11] Saleem, M., et al., *Experimental and theoretical study of highly porous lignocellulose assisted metal oxide photoelectrodes for dye-sensitized solar cells*. Arab. J. Chem., 2021. **14**(2): p. 102937.
[12] Wang, C., et al., *Near-infrared light induced in vivo photodynamic therapy of cancer based on upconversion nanoparticles*. Biomaterials, 2011. **32**(26): p. 6145–6154.
[13] Salvatore, D.A., et al., *Near-infrared–driven decomposition of metal precursors yields amorphous electrocatalytic films*. Sci. Adv., 2015. **1**(2): p. e1400215.
[14] Tsuji, Y., et al., *Near-infrared light-driven hydrogen evolution from water using a polypyridyl triruthenium photosensitizer*. Angew. Chem. Int. Ed, 2018. **57**(1): p. 208–212.
[15] Bhatt, M. and J.S. Lee, *Nanomaterials for photocatalytic hydrogen production: From theoretical perspectives*. RSC Adv., 2017. **7**: p. 34875–34885.
[16] Claridge, J.B., et al., *New catalysts for the conversion of methane to synthesis gas: molybdenum and tungsten carbide*. J. Catal., 1998. **180**(1): p. 85–100.
[17] K., A., *Photocatalysis and solar hydrogen production*. Pure Appl. Chem., 2007. **79**(11): p. 1917–1927.
[18] Li, Y. and J.Z. Zhang, *Hydrogen generation from photoelectrochemical water splitting based on nanomaterials*. Laser Photonics Rev., 2010. **4**(4): p. 517–528.
[19] Li, X., et al., *Photocatalytic hydrogen production over CdS nanomaterials: an interdisciplinary experiment for introducing undergraduate students to photocatalysis and analytical chemistry*. J. Chem. Educ., 2019. **96**(6): p. 1224–1229.
[20] Fallah, Z., et al., *Toxicity and remediation of pharmaceuticals and pesticides using metal oxides and carbon nanomaterials*. Chemosphere, 2021. **275**: p. 130055.
[21] Ding, Y., et al., *Insights into hydrotropic solubilization for hybrid ion redox flow batteries*. ACS Energy Letters, 2018. **3**(11): p. 2641–2648.
[22] Xiang, Q., F. Cheng, and D. Lang, *Hierarchical layered WS2/graphene-modified CdS nanorods for efficient photocatalytic hydrogen evolution*. ChemSusChem, 2016. **9**(9): p. 996–1002.
[23] Lang, D., T. Shen, and Q. Xiang, *Roles of MoS2 and graphene as cocatalysts in the enhanced visible-light photocatalytic H2 production activity of multiarmed CdS nanorods*. ChemCatChem, 2015. **7**(6): p. 943–951.
[24] Xiang, Q., et al., *Graphene-modified nanosized Ag3PO4 photocatalysts for enhanced visible-light photocatalytic activity and stability*. Appl. Catal. B., 2015. **162**: p. 196–203.
[25] Fu, J., et al., *Ultrathin 2D/2D WO3/g-C3N4 step-scheme H2-production photocatalyst*. Appl. Catal. B., 2019. **243**: p. 556–565.
[26] Liao, C.-H., C.-W. Huang, and J.C.S. Wu, *Hydrogen production from semiconductor-based photocatalysis via water splitting*. Catalysts, 2012. **2**(4): p. 490–516.
[27] Kashif, M., et al., *An experimental and DFT study on novel dyes incorporated with natural dyes on titanium dioxide (TiO2) towards solar cell application*. Appl. Phys. A, 2020. **126**(9): p. 716.
[28] Gordon, T.R., et al., *Nonaqueous synthesis of TiO2 nanocrystals using TiF4 to engineer morphology, oxygen vacancy concentration, and photocatalytic activity*. J. Am. Chem. Soc., 2012. **134**(15): p. 6751–6761.
[29] Duan, Y., et al., *TiO2 faceted nanocrystals on the nanofibers: Homojunction TiO2 based Z-scheme photocatalyst for air purification*. Appl. Surf. Sci., 2018. **456**: p. 817–826.
[30] Yu, J., et al., *Fabrication and enhanced visible-light photocatalytic activity of carbon self-doped TiO 2 sheets with exposed {001} facets*. J. Mater. Chem., 2011. **21**(4): p. 1049–1057.
[31] Wang, G., et al., *Hydrogen-treated TiO2 nanowire arrays for photoelectrochemical water splitting*. Nano Lett, 2011. **11**(7): p. 3026–33.
[32] Merlen, A., et al., *Surface enhanced Raman spectroscopy of organic molecules deposited on gold sputtered substrates*. Nanotechnology, 2009. **20**: p. 215705.

[33] Awazu, K., et al., *A plasmonic photocatalyst consisting of silver nanoparticles embedded in titanium dioxide*. J Am Chem Soc, 2008. **130**(5): p. 1676–80.

[34] Kudo, A., H. Kato, and S. Nakagawa, *Water splitting into H2 and O2 on new Sr2M2O7 (M = Nb and Ta) photocatalysts with layered perovskite structures: factors affecting the photocatalytic activity*. J. Phys. Chem., 2000. **104**: p. 571–575.

[35] Ganguly, P., et al., *2D Nanomaterials for photocatalytic hydrogen production*. ACS Energy Lett, 2019. **4**(7): p. 1687–1709.

[36] Ahmad, I., et al., *Lanthanum-zinc binary oxide nanocomposite with promising heterogeneous catalysis performance for the active conversion of 4-nitrophenol into 4-aminophenol*. Coatings, 2021. **11**: p. 537.

[37] Novoselov, K.S., et al., *Two-dimensional atomic crystals*. PNAS, 2005. **102**(30): p. 10451–10453.

[38] Castro Neto, A. and K. Novoselov, *Two-dimensional crystals: beyond graphene*. Opt. Mater. Express, 2011. **1**: p. 10–17.

[39] Bibi, S., et al., *Photocatalytic degradation of malachite green and methylene blue over reduced graphene oxide (rGO) based metal oxides (rGO-Fe3O4/TiO2) nanocomposite under UV-visible light irradiation*. J. Environ. Chem. Eng., 2021. **9**: p. 105580.

[40] Xia, D., et al., *Molybdenum and tungsten disulfides-based nanocomposite films for energy storage and conversion: A review*. Chem. Eng. J., 2018. **348**: p. 908–928.

[41] Coleman, J.N., et al., *Two-dimensional nanosheets produced by liquid exfoliation of layered materials*. Science, 2011. **331**(6017): p. 568–71.

[42] Wang, X., et al., *Atomic-scale clarification of structural transition of MoS 2 upon sodium intercalation*. ACS Nano, 2014. **8**.

[43] Reina, A., et al., *Large area, few-layer graphene films on arbitrary substrates by chemical vapor deposition*. Nano Letters, 2009. **9**(1): p. 30–35.

[44] Shi, Y., H. Li, and L.-J. Li, *Recent advances in controlled synthesis of two-dimensional transition metal dichalcogenides via vapour deposition techniques*. Chem. Soc. Rev., 2015. **44**(9): p. 2744–2756.

[45] Bertolazzi, S., et al., *Molecular chemistry approaches for tuning the properties of two-dimensional transition metal dichalcogenides*. Chem. Soc. Rev., 2018. **47**(17): p. 6345–6888.

[46] Yuan, H., et al., *Modulating the external facets of functional nanocrystals enabled by two-dimensional oxide crystal templates*. ACS catalysis, 2017. **7**(10): p. 6858–6863.

[47] Wen, J., et al., *A review on g-C3N4-based photocatalysts*. Appl. Surf. Sci., 2017. **391**: p. 72–123.

[48] Mamba, G. and A.K. Mishra, *Graphitic carbon nitride (g-C3N4) nanocomposites: A new and exciting generation of visible light driven photocatalysts for environmental pollution remediation*. Appl. Catal. B., 2016. **198**: p. 347–377.

[49] Kashif, M., et al., *Effect of potassium permanganate on morphological, structural and electro-optical properties of graphene oxide thin films*. Arab. J. Chem., 2021. **14**(2): p. 102953.

[50] Wen, J., et al., *Enhanced visible-light H2 evolution of g-C3N4 photocatalysts via the synergetic effect of amorphous NiS and cheap metal-free carbon black nanoparticles as co-catalysts*. Appl. Surf. Sci., 2015. **358**: p. 204–212.

[51] Lu, Y., et al., *Exfoliated carbon nitride nanosheet decorated with NiS as efficient noble-metal-free visible-light-driven photocatalyst for hydrogen evolution*. Phys. Chem. Chem. Phys., 2015. **17**.

[52] Liu, M., et al., *Photocatalytic hydrogen production using twinned nanocrystals and an unanchored NiSx co-catalyst*. Nat. Energy, 2016. **1**.

[53] Zhong, Y., et al., *Earth-abundant NiS Co-catalysts modified metal-free mpg-C3N4/CNTs nanocomposites for highly efficient visible-light photocatalytic H2 evolution*. Dalton Trans., 2015. **44**.

[54] Seoane, B., et al., *Multi-scale crystal engineering of metal organic frameworks*. Coord. Chem. Rev., 2016. **307**: p. 147–187.

[55] Hu, S., et al., *Construction of g-C3N4/Zn0.11Sn0.12Cd0.88S1.12 hybrid heterojunction catalyst with outstanding nitrogen photofixation performance induced by sulfur vacancies*. ACS Sustain. Chem. Eng., 2016. **4**(4): p. 2269–2278.

[56] Lu, A.Y., et al., *High-sulfur-vacancy amorphous molybdenum sulfide as a high current electrocatalyst in hydrogen evolution*. Small, 2016. **12**(40): p. 5530–5537.

[57] Huang, Y., et al., *Low concentration nitric acid facilitate rapid electroŽ hole separation in vacancy-rich bismuth oxyiodide for photo-thermo-synergistic oxidation of formaldehyde*. Appl. Catal. B., 2017. **218**: p. 700–708.

[58] Das, D. and T.N. Veziroğlu, *Hydrogen production by biological processes: a survey of literature*. Int. J. Hydrog. Energy, 2001. **26**(1): p. 13–28.

[59] Sakurai, H., et al., *Photobiological hydrogen production: Bioenergetics and challenges for its practical application*. J. Photochem. Photobiol. C: Photochem. Rev., 2013. **17**: p. 1–25.

[60] Jose, V., *Photobiological hydrogen as a renewable fuel*. The holistic approach to environment, 2021. **11**: p. 67–71.

[61] Asada, Y. and J. Miyake, *Photobiological hydrogen production*. J Biosci Bioeng, 1999. **88**(1): p. 1–6.

[62] Sayama, K. and H. Arakawa, *Effect of carbonate salt addition on the photocatalytic decomposition of liquid water over Pt–TiO2 catalyst*. J. Chem. Soc. Faraday Trans., 1997. **93**(8): p. 1647–1654.

[63] Sayama, K. and H. Arakawa, *Photocatalytic decomposition of water and photocatalytic reduction of carbon dioxide over zirconia catalyst*. J. Phys. Chem. A, 1993. **97**(3): p. 531–533.

[64] Kato, H. and A. Kudo, *New tantalate photocatalysts for water decomposition into H2 and O2*. Chem. Phys. Lett., 1998. **295**: p. 487–492.

[65] Shimizu, K., et al., *Photocatalytic water splitting on Ni-intercalated Ruddlesden–Popper tantalate H2La2/3Ta2O7*. Chem. Mater., 2005. **17**(20): p. 5161–5166.

[66] Higashi, M., et al., *Photocatalytic overall water splitting under visible light using ATaO2N (A = Ca, Sr, Ba) and WO3 in a* IO_3^-/I^- *shuttle redox mediated system*. Chem. Mater., 2009. **21**(8): p. 1543–1549.

[67] Kato, H., et al., *Construction of Z-scheme type heterogeneous photocatalysis systems for water splitting into H2 and O2 under visible light irradiation*. Chem Lett, 2004. **33**: p. 1348–1349.

[68] Lo, C.-C., et al., *Novel twin reactor for separate evolution of hydrogen and oxygen in photocatalytic water splitting*. Int. J. Hydrog. Energy, 2010. **35**(4): p. 1523–1529.

[69] Li, C., et al., *Visible light driven hydrogen evolution by photocatalytic reforming of lignin and lactic acid using one-dimensional NiS/CdS nanostructures*. Appl. Catal. B., 2018. **227**: p. 229–239.

[70] Li, C., et al., *Hydrogen generation by photocatalytic reforming of glucose with heterostructured CdS/MoS2 composites under visible light irradiation*. Int. J. Hydrog. Energy, 2017. **42**(27): p. 16968–16978.

[71] Speltini, A., et al., *Improved photocatalytic H2 production assisted by aqueous glucose biomass by oxidized g-C3N4*. Int. J. Hydrog. Energy 2018. **43**(32): p. 14925–14933.

Hina Sharif, Awais Ahmad, Rafael Luque, Mabkhoot Alsaiari, Farid A. Harraz

9 Recent advancement and development in oxygen evaluation using biomaterials

Abstract: The process of tissue engineering and chemotherapy has offered significant strategies for enhancing the quality of life and combating contagious diseases. However, there are several limitations associated with these clinical therapies, including limited oxygen supply to the tissue constructs, which can, in turn, cause hypoxia and tissue resistance. Moreover, oxygen is also required as a significant signaling tool for the control of stem cell differentiation within the tissue engineering scaffolds. Several techniques have been developed and investigated for the process of oxygen generation. The use of biomaterials can be beneficial for the process of oxygen generation through their photocatalytic activity. This overview provides detail of the approaches of oxygen generation and the use of biomaterials in this technology.

Keywords: Biomaterials, ORR, oxygen generation

9.1 Introduction

Currently, the world is affected by several contagious and infectious diseases that can be treated with tissue transplantation and/or oxygen [1]. In the practical process of tissue transplantation, the concerns of tissue functionality and cell survival are critical, and the main problem is the insufficient oxygen supply to the cells and tissues during the initial stages of implantation. Therefore, oxygen supply is one of major elements for the survival of cells and tissues [2, 3]. Because of the insufficient oxygen supply during the stages of tissue transplantation, hypoxia may occur in the deeper regions of the tissues [4]. Several strategies have been designed to overcome this limitation over the last decade, such as using synthetic carriers, fabrication of oxygen-generating scaffolds, enhancing the vascularization process and development of oxygen-generating microsystems [5–7]. Hypoxia is the major drawback in the anticancer therapy of tumorous cells and tissue engineering, and hypoxia is the major concern for tissue regeneration, as it causes resistance to cancer cells [8]. To overcome these problems, several approaches have been made for restoring adequate oxygen tension at the defected site for enhancing therapeutic outcomes, such as the use of oxygen carriers, direct oxygen transplantation and oxygen-generating materials [9, 10].

The process of oxygenation can be achieved through three ways: hyperbaric oxygen therapy (HBO), oxygen-carrying biomaterials (OCB) and oxygen-generating

https://doi.org/10.1515/9783110768749-009

biomaterials (OBGs). The HBO technique was historically used for the treatment of decompression sickness in deep-sea divers since 1920s; however, recently this technique has been investigated for the tissue engineering process [11]. Among OCBs, perfluorocarbon (PFC) has been widely investigated for the applications of tissue engineering, while among OGBs, peroxides and their salts are commonly employed for the process of oxygenation [12].

Oxygen-releasing biomaterials (ORBs) is one of the successful attempts in providing the prolonged and sufficient oxygen supply for the engineered tissues in both in vitro and in vivo in clinical practice [13, 14]. These biomaterials work by releasing oxygen by the diffusion of adsorbed or entrapped oxygen and/or chemical production of oxygen. OGBs is the category of ORBs that are designed for providing a moderate release of oxygen to cells and tissues [15]. The moderate and gradual release of oxygen is the essential parameter for the organ, particularly providing sufficient time for the maturation and neovascularization of the engineered tissues [13]. These materials are usually incorporated into the engineered tissues constructs in the form of nanofibers, scaffolds and microsphere. Fluorinated materials, calcium peroxide, hydrogen peroxide, magnesium peroxide and sodium percarbonate are some examples of ORBs [6, 16]. Encapsulation is another modification of biomaterials using hydrophobic polymers that can also further slow down the oxygen release. Moreover, the chemical composition of OGBs can also affect the process of oxygenation [16].

9.2 Approaches of oxygen generation

9.2.1 Hyperbaric oxygen therapy (HBO_2)

This approach possesses much potential for the technique of tissue engineering and wound healing. In this technique, patients usually breathe 100% oxygen in their chamber at an elevated atmospheric pressure, which increases the oxygen level in the blood and tissues. The major drawback of this approach is the generation of ROS, which is considered to be toxic to cells and can cause systemic impairment to the central nervous system and the pulmonary tract. However, these outcomes are dependent upon the concentration of generated ROS and the duration of the therapy [11, 17]. These potential drawbacks of the HBO_2 technique have greatly influenced the investigation and research for their applications in the tissue engineering of the cartilage, ligament, bone and the oral mucosa [18–20]. Table 9.1 explores the pros and cons of the HBO_2 technique for oxygen generation. However, the process of HBO_2 is not effective for many reasons. The major method of its delivery is through the arterial blood, which may cause integration, and moreover, this technique does not provide a sustained oxygen level and it also needs specialized equipment [11].

Table 9.1: Pros and cons of HBO_2 therapy.

Pros	Cons
– Simple and easy technique – The process of oxygen delivery can be sustained	– Requires daily treatment – Oxygen is often flammable at high concentration – Seizures, reversible myopia and pulmonary damage may occur – The delivery of oxygen is sustained by the process of diffusion

9.2.2 Oxygen-carrying biomaterials (OCBs)

PFCs can be used as OCBs that consist of fluorinated carbon chains exhibiting several properties, which make them suitable oxygen carriers for biomedical applications. The bonds between carbon and fluorine are responsible for the inertness of PFCs and also make them biocompatible with living cells [21]. PFCs show both lipophilic and hydrophilic character that contribute to the structural stability of PFCs with living tissues [22]. All these characteristics of PFCs make them a suitable choice for storing and releasing oxygen on metabolic demands [23]. Several attempts and investigations have been made clinically on the use of PFC emulsions as artificial blood substituents. Oxygent, a PFC emulsion having 58% perflubron is one of the successfully employed artificial blood substituents [24]. In spite of several investigations, no PFC has been currently approved by the FDA.

PFCs can also be incorporated into the matrices of biomaterials, which can in turn be used in several applications. The incorporation of PFC into biomaterials requires fabrication in the encapsulating cells [25, 26]. Chin et al. encapsulated the perflluorooctylbromide emulsions in the hydrogel beads, which in turn enhanced the cells viability and the metabolic activity in the tissue-engineered liver constructs [27]. Instead of hydrogels, other agents like solvent-casted polycarbonate urethane, along with fluorinated zeolite microparticles, can also be used to overcome the limitation of hydrogels with PFCs [28].

Peroxides can also be used as OCB, as these substances work by generating oxygen, instead of delivering it [29]. Inorganic peroxide, such as calcium peroxide, can also be used for the process of oxygen generation due to its longer release time [6].

9.2.3 Oxygen-generating biomaterials (OGBs)

Calcium peroxide, one of the OGBs, possesses the capability of producing ROS as a result of the oxygen decomposition process. Through building up of H_2O_2, the production of free radicals is remarkably enhanced, which can decrease cell viability. Hence, catalase can be incorporated as an antioxidant, which can enhance the degradation of hydrogen peroxide. For the in vivo utilization of OGBs, engineered tissues might be given catalase to lessen hydrogen peroxide's harmful effects when it builds up (Table 9.2) [6].

Table 9.2: List of oxygen-generating biomaterials (OGBs) used in in vitro investigation.

OGBs/ORBs	Advantages	Limitations	References
Hydrogen peroxide (H_2O_2)	Moderate release of oxygen and also reduces the diffusion of oxygen around the scaffold	Less oxygenation consistency, needs modification for in vivo incorporation	[30]
Calcium peroxide	Sustained release of oxygen for about 42 days reduces the effects of oxidative stress on cells	Less oxygenation consistency and cytotoxic effects on human cells; needs further modification for in vivo utilization	[15]
Perfluorocarbons	Sustained release of oxygen	To be included in a suitable scaffold for large tissue implantations.	[31, 32]

9.2.4 Mechanism of oxygen generation

Oxygen rate generation and release from biomaterials are significant factors that can directly influence new tissues formation. This mechanism of oxygen release should be adjusted and moderated for the sustained release of oxygen depending on the needs of the target tissues. This is due to fact that a burst and/or prolonged oxygen release cannot provide a significant amount of oxygen to the living tissues, reducing the potential of cellular functions. To address these problems, a number of factors including pH, temperature, type of medium and the amount of peroxide should be considered [33].

Solid inorganic peroxides have been investigated for promising OGBs for a large number of tissue engineering applications. Calcium peroxide, sodium percarbonate and magnesium peroxide have been reported to be beneficial for tissue engineering, with calcium peroxide being the most propitious OGB [34, 35]. As compared to calcium peroxide, magnesium peroxide shows the slowest oxygen generation due to its slower reaction rate and lower solubility. Several strategies have been successfully adopted to enhance the rate of oxygen generation via these OGBs. For instance, encapsulation in hydrophobic materials can slow the oxygen release rate, while encapsulation in hydrophilic materials can enhance the diffusion rate [13, 34].

9.2.5 Photosynthetic microorganisms for potent oxygen generators

The use of photosynthetic microorganisms has recently become a continuous source of oxygenation as an unconventional promising prospect [36, 37]. The mechanism involved is the fixation of carbon dioxide and water, and the generation of oxygen as a byproduct through light-sensitive chlorophyll molecules [37]. Thus, this system of oxygen generation can be effectively utilized for overcoming the process of hypoxia in tissue engineering and in the chemotherapy of cancerous cells [8]. The main concern of utilizing this system is the delivery of microbes into the mammalian systems. The three major approaches used for this purpose include direct treatment, encapsulation into the matrix and adhesion on the biomaterial constructs. In the direct treatment method, the microbes are delivered in the form of suspension through intravenous injection at the specific target cells [38]. The major drawback of using this method is less specificity and rapid clearance through the liver and spleen. The delivery approach of the adhesion of microbes on biomaterial constructs involves the resuspension of the biomaterial platform into the microbial cell suspension, which allow the cells to adhere firmly toward the substrate surface [39, 40]. This method confers a site-specific action, providing a high concentration. Nonetheless, the microbial cells can elicit an immune response into the host cells. Hence, another approach can be adopted, which involves the encapsulation of microbial cells in three-dimensional matrices that offers site-specific delivery and less immune interactions [41–43]. Tables 9.3 and 9.4 summarize the study of photosynthetic microorganism utilized biomaterials for tissue engineering and anticancer therapy, respectively.

Algae can also be used successfully for the constant supply of oxygen into the constructs of tissue engineering due to its photosynthetic ability [44]. Algae can be used alone or with fabrication, which further increases the process of oxygenation. In a recent research, it was found that algae can also be genetically engineered for the secretion of vascular endothelial growth factor, which can in turn enhance the vascularization and oxygen generation within the tissue constructs [45].

Table 9.3: Photosynthetic microorganism utilized biomaterials for tissue engineering.

S. no.	Microorganisms	Method of delivery	In vitro study	In vivo study	References
1	*Chlorella sorokiniana*	Direct encapsulation	Mouse pancreatic islet cells		[46]
2	*C. reinhardtii*	Loading on scaffolds		Mouse with full skin defect	[47]
3	*C. reinhardtii*	Resuspension	SaOS-2 human osteosarcoma cells		[48]
4	*Spirulina platensis*	Resuspension	Epithelial cells	Full-skin defect mouse	[49]
5	*S. elongatus*	Direct intravenous injection	Neonatal rat cardiomyocytes (CMs)	Myocardial infarcted rat	[50]
6	*S. elongatus*	Direct encapsulation	Human skin fibroblasts	Full-skin chronic defect diabetic mouse	[51]

Table 9.4: Photosynthetic microorganism utilized biomaterials for anticancers.

S. no.	Microorganisms	In vitro study	In vivo study	References
1	*Chlorella* species	CT26 cells	Mouse model bearing CT26 tumor	[52]
2	*Chlorella pyrenoidosa*	4T1 cells	Mouse model bearing 4T1 tumor	[53]
3	*C. vulgaris*	4T1 cells	Mouse model bearing 4T1 tumor	[54]
4	*S. elongatus*	MGC-803 cells	Mouse models bearing B16-F10 tumor	[38]

9.3 Application of OGBs in tissue engineering

In the recent years, several investigations and developments have been carried out on the incorporation of OGBs into the tissue-engineered constructs through resultant devices for in vivo use [16, 55]. For instance, several investigations have been conducted for the provision of sufficient oxygen levels to islet cells for the treatment of diabetes mellitus (DM) [56, 57]. Pedraza et al. formulated a unique construct using polydimethylsiloxane-encapsulated calcium peroxide (PDMS-CaO_2), which was found to be novel in enhancing the duration of oxygen release for in vitro studies. However, this construct needs further research for evaluating its efficacy and behavior in vivo [56].

Similarly, there is a problem of insufficient oxygen level in the transplanted tissues of a dermal wound because of the loss of vascular network during injuries.

Various attempts have been made on incorporating the PFCs into methacrylamide chitosan (MACF) for developing a novel injectable hydrogel that is found to be efficacious in repetitive oxygen absorption and emission for the healing process [32]. Apart from this, hydrogel made of fluorinated MACF is investigated to be conveniently refilled with oxygen. It has also been investigated that increasing the concentration of PFC with the construct showed positive effects on the metabolic activity of fibroblasts. However, this construct requires further investigation in wound healing for in vivo applications [13, 58]. Moreover, several futuristic investigations are also needed to extend the oxygen production without affecting the process of cell migration and/or cell adhesion adversely [59].

Muscle tissues need high oxygen levels and, hence, deficiency of oxygen after transplantation may lead to ischemic conditions and necrosis due to poor vascular networks [60]. Therefore, a moderate supply of oxygen during the early stages of tissue transplantation is very essential for preventing the effect of hypoxia. The application of OGBs can be used as a substitute method for ensuring an adequate supply of oxygen [61]. PLGA microsphere, enclosed in hydrogen peroxide, encircled by alignate is one of the successful attempts of using biomaterials for the oxygenation of muscle tissue after tissue transplantation [62].

9.4 Conclusion

Oxygenation therapy, through photosynthesis, using biomaterials holds a massive therapeutic potential for regulating the process of chemotherapy and tissue engineering. This approach of oxygenation shows promising applications for the repairing and regeneration process of the defected cells after tissue engineering, and for reversing the process of hypoxia and resistance to anticancer agents. Despite this exciting technology, there are several drawbacks and limitations associated with it. Through continuous efforts and research, this field of technology can be used for better oxygenation in tissue engineering and chemotherapy in the future.

References

[1] Huh, J.-S., *Functionalized biomaterials – oxygen releasing scaffolds*. J. Biotechnol. Biomater., 2015. **05**.

[2] Place, E.S., N.D. Evans, and M.M. Stevens, *Complexity in biomaterials for tissue engineering*. Nat Mater, 2009. **8**(6): p. 457–70.

[3] Phelps, E.A. and A.J. García, *Engineering more than a cell: vascularization strategies in tissue engineering*. Curr Opin Biotechnol, 2010. **21**(5): p. 704–9.

[4] Malda, J., et al., *The effect of PEGT/PBT scaffold architecture on oxygen gradients in tissue engineered cartilaginous constructs*. Biomaterials, 2004. **25**(26): p. 5773–80.

[5] Iyer, R.K., et al., *Synthetic oxygen carriers in cardiac tissue engineering*. Artif Cells Blood Substit Immobil Biotechnol, 2007. **35**(1): p. 135–48.
[6] Oh, S.H., et al., *Oxygen generating scaffolds for enhancing engineered tissue survival*. Biomaterials, 2009. **30**(5): p. 757–62.
[7] Ng, S.M., et al., *Novel microencapsulation of potential drugs with low molecular weight and high hydrophilicity: hydrogen peroxide as a candidate compound*. Int J Pharm, 2010. **384** (1–2): p. 120–7.
[8] Agarwal, T., M. Costantini, and T.K. Maiti, *Recent advances in tissue engineering and anticancer modalities with photosynthetic microorganisms as potent oxygen generators*. Adv. Biomed. Eng., 2021. **1**: p. 100005.
[9] Huang, C.-C., et al., *An Implantable Depot That Can Generate Oxygen in Situ for Overcoming Hypoxia-Induced Resistance to Anticancer Drugs in Chemotherapy*. JACS, 2016. **138**(16): p. 5222–5225.
[10] Agarwal, T., et al., *Oxygen releasing materials: Towards addressing the hypoxia-related issues in tissue engineering*. Mater. Sci. Eng. C 2021. **122**: p. 111896.
[11] Farris, A.L., A.N. Rindone, and W.L. Grayson, *Oxygen Delivering Biomaterials for Tissue Engineering*. J Mater Chem B, 2016. **4**(20): p. 3422–3432.
[12] Ishaug, S.L., et al., *Bone formation by three-dimensional stromal osteoblast culture in biodegradable polymer scaffolds*. J Biomed Mater Res, 1997. **36**(1): p. 17–28.
[13] Gholipourmalekabadi, M., et al., *Oxygen-generating biomaterials: a new, viable paradigm for tissue engineering?* Trends in Biotechnology, 2016. **34**(12): p. 1010–1021.
[14] Lee, H.-Y., et al., *Controlling oxygen release from hollow microparticles for prolonged cell survival under hypoxic environment*. Biomaterials, 2015. **53**: p. 583–591.
[15] Li, Z., X. Guo, and J. Guan, *An oxygen release system to augment cardiac progenitor cell survival and differentiation under hypoxic condition*. Biomaterials, 2012. **33**(25): p. 5914–5923.
[16] Harrison, B.S., et al., *Oxygen producing biomaterials for tissue regeneration*. Biomaterials, 2007. **28**(31): p. 4628–4634.
[17] Clark, J.M., *The toxicity of oxygen*. Am Rev Respir Dis, 1974. **110**(6 Pt 2): p. 40–50.
[18] Ishii, Y., et al., *Effects of different exposures of hyperbaric oxygen on ligament healing in rats*. J Orthop Res, 2002. **20**(2): p. 353–6.
[19] Cherng, J.H., et al., *The effect of hyperbaric oxygen and air on cartilage tissue engineering*. Ann Plast Surg, 2012. **69**(6): p. 650–5.
[20] Pedersen, T.O., et al., *Hyperbaric oxygen stimulates vascularization and bone formation in rat calvarial defects*. Int J Oral Maxillofac Surg, 2013. **42**(7): p. 907–14.
[21] Riess, J.G. and M.P. Krafft, *Fluorinated materials for in vivo oxygen transport (blood substitutes), diagnosis and drug delivery*. Biomaterials, 1998. **19**(16): p. 1529–39.
[22] Riess, J.G., *Understanding the fundamentals of perfluorocarbons and perfluorocarbon emulsions relevant to in vivo oxygen delivery*. Artif Cells Blood Substit Immobil Biotechnol, 2005. **33**(1): p. 47–63.
[23] Squires, J.E., *Artificial blood*. Science, 2002. **295**(5557): p. 1002–1005.
[24] Cohn, C.S. and M.M. Cushing, *Oxygen therapeutics: perfluorocarbons and blood substitute safety*. Crit Care Clin, 2009. **25**(2): p. 399–414, Table of Contents.
[25] Ma, T., et al., *The effect of synthetic oxygen carrier-enriched fibrin hydrogel on Schwann cells under hypoxia condition in vitro*. Biomaterials, 2013. **34**(38): p. 10016–27.
[26] White, J.C., et al., *Addition of perfluorocarbons to alginate hydrogels significantly impacts molecular transport and fracture stress*. J Biomed Mater Res A, 2013. **101**(2): p. 438–46.
[27] Chin, K., et al., *Hydrogel-perfluorocarbon composite scaffold promotes oxygen transport to immobilized cells*. Biotechnol Prog, 2008. **24**(2): p. 358–66.

[28] Seifu, D.G., T.T. Isimjan, and K. Mequanint, *Tissue engineering scaffolds containing embedded fluorinated-zeolite oxygen vectors*. Acta Biomater, 2011. **7**(10): p. 3670–8.

[29] Harrison, B.S., et al., *Oxygen producing biomaterials for tissue regeneration*. Biomaterials, 2007. **28**(31): p. 4628–34.

[30] Abdi, S.I.H., S.M. Ng, and J.O. Lim, *An enzyme-modulated oxygen-producing micro-system for regenerative therapeutics*. Int. J. Pharm., 2011. **409**(1): p. 203–205.

[31] Khattak, S.F., et al., *Enhancing oxygen tension and cellular function in alginate cell encapsulation devices through the use of perfluorocarbons*. Biotechnol. Bioeng., 2007. **96**(1): p. 156–166.

[32] Wijekoon, A., N. Fountas-Davis, and N.D. Leipzig, *Fluorinated methacrylamide chitosan hydrogel systems as adaptable oxygen carriers for wound healing*. Acta Biomateralia, 2013. **9**(3): p. 5653–5664.

[33] Soleymani, M., A. Moheb, and D. Babakhani, *Hydrogen Peroxide Decomposition over Nanosized La1–XCaXMnO3 (0 ≤ X ≤ 0.6) Perovskite Oxides*. Chem Eng Technol 2011. **34**(1): p. 49–55.

[34] Fraker, C.A., A.J. Mendez, and C.L. Stabler, *Complementary Methods for the Determination of Dissolved Oxygen Content in Perfluorocarbon Emulsions and Other Solutions*. J. Phys. Chem. B, 2011. **115**(35): p. 10547–10552.

[35] Chandra, P.K., et al., *Peroxide-based oxygen generating topical wound dressing for enhancing healing of dermal wounds*. Wound Repair and Regeneration, 2015. **23**(5): p. 830–841.

[36] Hu, H., X. Qian, and Y. Chen, *Microalgae-enabled photosynthetic alleviation of tumor hypoxia for enhanced nanotherapies*. Sci. Bull., 2020. **65**(22): p. 1869–1871.

[37] Chávez, M.N., et al., *Photosymbiosis for Biomedical Applications*. Front. bioeng. biotechnol., 2020. **8**.

[38] Zhang, Y., et al., *Cyanobacteria-based near-infrared light-excited self-supplying oxygen system for enhanced photodynamic therapy of hypoxic tumors*. Nano Research, 2021. **14**(3): p. 667–673.

[39] Centeno-Cerdas, C., et al., *Development of photosynthetic sutures for the local delivery of oxygen and recombinant growth factors in wounds*. Acta Biomaterialia, 2018. **81**: p. 184–194.

[40] Hopfner, U., et al., *Development of photosynthetic biomaterials for in vitro tissue engineering*. Acta Biomaterialia, 2014. **10**(6): p. 2712–2717.

[41] Lee, C., et al., *Chlorella-gold nanorods hydrogels generating photosynthesis-derived oxygen and mild heat for the treatment of hypoxic breast cancer*. J. Control. Decis., 2019. **294**: p. 77–90.

[42] Zhou, T.-J., et al., *Light triggered oxygen-affording engines for repeated hypoxia-resistant photodynamic therapy*. J Contr Rel, 2019. **307**: p. 44–54.

[43] Maharjan, S., et al., *Symbiotic Photosynthetic Oxygenation within 3D-Bioprinted Vascularized Tissues*. Matter, 2021. **4**(1): p. 217–240.

[44] Hopfner, U., et al., *Development of photosynthetic biomaterials for in vitro tissue engineering*. Acta Biomater, 2014. **10**(6): p. 2712–7.

[45] Chávez, M.N., et al., *Towards autotrophic tissue engineering: Photosynthetic gene therapy for regeneration*. Biomaterials, 2016. **75**: p. 25–36.

[46] Bloch, K., et al., *Photosynthetic oxygen generator for bioartificial pancreas*. Tissue Eng, 2006. **12**(2): p. 337–44.

[47] Schenck, T.L., et al., *Photosynthetic biomaterials: A pathway towards autotrophic tissue engineering*. Acta Biomaterialia, 2015. **15**: p. 39–47.

[48] Lode, A., et al., *Green bioprinting: Fabrication of photosynthetic algae-laden hydrogel scaffolds for biotechnological and medical applications*. Eng. Life Sci., 2015. **15**(2): p. 177–183.
[49] Li, W., et al., *A Bioactive Living Hydrogel: Photosynthetic Bacteria Mediated Hypoxia Elimination and Bacteria-Killing to Promote Infected Wound Healing*. Adv. Ther., 2020. **4**: p. 2000107.
[50] C., J.E., et al., *An innovative biologic system for photon-powered myocardium in the ischemic heart*. Science Advances, 2017. **3**(6): p. e1603078.
[51] Chen, H., et al., *Dissolved oxygen from microalgae-gel patch promotes chronic wound healing in diabetes*. Science Advances, 2020. **6**(20):p. eaba4311.
[52] Wang, H., et al., *Light-controlled oxygen production and collection for sustainable photodynamic therapy in tumor hypoxia*. Biomaterials, 2021. **269**: p. 120621.
[53] Zhou, T.-J., et al., *Light triggered oxygen-affording engines for repeated hypoxia-resistant photodynamic therapy*. J. Controlled Release, 2019. **307**: p. 44–54.
[54] Lee, C., et al., *Chlorella-gold nanorods hydrogels generating photosynthesis-derived oxygen and mild heat for the treatment of hypoxic breast cancer*. J. Controlled Release, 2019. **294**: p. 77–90.
[55] Ward, C.L., et al., *Oxygen generating biomaterials preserve skeletal muscle homeostasis under hypoxic and ischemic conditions*. PLoS One, 2013. **8**(8): p. e72485.
[56] Pedraza, E., et al., *Preventing hypoxia-induced cell death in beta cells and islets via hydrolytically activated, oxygen-generating biomaterials*. PNAS, 2012. **109**(11): p. 4245–4250.
[57] Colton, C.K., *Oxygen supply to encapsulated therapeutic cells*. Adv. Drug Deliv. Rev., 2014. **67–68**: p. 93–110.
[58] Wilcox, J.R. and D.S. Covington, *Wound healing and hyperbaric oxygen therapy physiology: oxidative damage and antioxidant imbalance*, in *Oxidative Stress and Antioxidant Protection*. 2016. p. 347–356.
[59] Zellner, S., R. Manabat, and D.F. Roe, *A dissolved oxygen dressing: a pilot study in an ischemic skin flap model*. J Int Med Res, 2015. **43**(1): p. 93–103.
[60] Abdi, S.I.H., et al., *Controlled release of oxygen from PLGA-alginate layered matrix and its in vitro characterization on the viability of muscle cells under hypoxic environment*. J Tissue Eng Regen Med, 2013. **10**(3): p. 131–138.
[61] Seifu, D.G., T.T. Isimjan, and K. Mequanint, *Tissue engineering scaffolds containing embedded fluorinated-zeolite oxygen vectors*. Acta Biomaterialia, 2011. **7**(10): p. 3670–3678.
[62] Abdi, S.I.H., et al., *Controlled release of oxygen from PLGA-alginate layered matrix and its in vitro characterization on the viability of muscle cells under hypoxic environment*. J Tissue Eng Regen Med 2013. **10**(3): p. 131–138.

Awais Ahmad, Hina Sharif, Rafael Luque, Mabkhoot Alsaiari,
Farid A. Harraz

10 Light-driven photocatalysis using biomaterials for biomedical applications

Abstract: Photocatalytic reactions using biomaterials is the emerging technology for the development of non-invasive and controllable medical devices in several biomedical applications such as photodynamic therapy, photothermal, drug delivery, regenerative medicines, disinfection and pollutant degradations. Carbon-based nanomaterials, graphene-based nanomaterials and gold nanoparticles are largely used in light-guided medical devices and biomedical applications. This chapter highlights the importance of such biomaterials in the field of medical and advanced therapies, which can offer huge promise in the upcoming treatment strategies and for meeting the future energy demands.

Keywords: Photocatalysis, CNTs, graphene, gold NPs

10.1 Introduction

Industrialization and urbanization have led to the current world, and to the consequential environmental and energy crises. To overcome these problems, an effective utilization of solar light has driven many interests in the fields of research and development, as this source of energy that reaches the Earth from the Sun in an hour exceeds the overall energy consumed by man in 1 year [1]. After the successful photocatalytic splitting of water using TiO_2 by Honda-Fujishima in the 1970s, the process of photocatalysis has gained huge interest in several research fields [2, 3]. The process of photocatalysis is rather complicated, which involves the absorption of photon to generate the electron–hole pairs and these electron–hole pairs are capable of initiating redox reactions [4, 5]. Based on such mechanisms, light can be successfully used in modern medicines and therapies for the treatment of several disorders, such as melanoma, acne, skin disorders and depression [6]. Light is an exceptionally captivating source of energy as compared to magnetic and/or thermal energy and as an external force for controlling the stimuli-responsive biomaterials. This source of energy offers several advantages in terms of biomedical applications – light can be confined in a predefined area through modern optical instruments, the stimulation characteristics can be refined by intensity or duration of light exposure, different biomaterials can independently control the specific wavelength of light to generate different bioinductive signals for managing

https://doi.org/10.1515/9783110768749-010

the cell–biomaterial interactions and, more importantly, the light source is a harmless choice physiologically for the control of biomaterials in the physiological environment [6–8].

Photocatalysts can be effectively used for the effective utilization of solar energy. The limitation of the use of photocatalysis is a need for a continuous solar source to perform redox reactions, and that limits their wide use at night [9]. To overcome this limitation, round-the-clock photocatalyst (RTCP) or all-day-active photocatalysts are in the development and research stage to perform the catalytic processes throughout the day [10]. The research interest of today era has greatly shifted toward RTCP, also known as memory catalysis (MC), whose mechanisms of action are quite diverse and different. For this purpose, energy storage substances (ESS) are also required that are capable of initiating the catalysis in the dark [11]. ESS are capable of storing the photo catalytic-generated electrons from the photocatalyst under solar light and they can then degrade them in the dark [1]. Hence, the performance of RTCP is based on the combination of semiconductors and ESS. The development and advancement in photocatalysts, especially RTCP, finds significant applications, including degrading the organic pollutants, disinfection, removing heavy metal ions in water treatment and hydrogen and oxygen production [12–14].

By the effective usage of these photocatalytic substances, several biomedical applications have emerged that can use the proper source of light such as for the removal of heavy metals from water, artificial photosynthesis, disinfection, pollutant degradation, photothermal therapy (PTT) and PDT, cells optical modulation, drug delivery system and light-guided nanovehicles [1, 6]. These offer interesting options to address medical deficiencies and achieving targeted therapy to excel the treatment strategies.

10.2 Biomedical applications

10.2.1 Removing the ions of heavy metals

TiO_2–WO_3 systems have effectively reduced the ions of heavy metals (Hg, Ag and Cr), even in the absence of light. They can store the electrons under ultraviolet light and then these stored electrons can be released for reducing the heavy metal ions [15]. The fact is that heavy metals are good electron acceptors and, consequently, they can be decreased by the TiO_2–WO_3's trapped electrons. The presence of oxygen molecules can hinder the electron storage; however, the electrons that react with oxygen can generate ROS that can further be used in the degradation process. Due to this fact, the charging process needs the absence of oxygen, while the discharging process needs the presence of oxygen to generate ROS for environmental detoxification

[1, 16]. Keeping in view the environmental protection, the use of photocatalytic removal of heavy metals, not only at day time but also at night time, is an interesting strategy and is in need of future development [1].

10.2.2 Pollutant degradation

Phenol, formate, alcohols and aldehydes can be oxidized using the TiO_2–$Ni(OH)_2$ bilyer film, both under light and in dark. Moreover, TiON/PdO can also be used for the degradation of dyes [11, 17]. Based on recent research, g-C_3N_4/CNTs/graphene was also found to be effective as a photocatalyst for the degradation of phenol [18].

10.2.3 Disinfection

Biomaterials such as TiO_2–WO_3 can also be used as effective alternatives for bactericidal disinfections,. The discovery of this photocatalyst has been found to be a fundamental mechanism of the photocatalystic activity under dark conditions for antimicrobial applications. The photocatalytic films of TiO_2–WO_3 possesses the capability of being charged under UV light, and thus can display bactericidal effect in the dark for about 6 h against *Escherichia coli* (*E. coli*) [13]. The mechanism of disinfection of semiconductors, based on their mechanism of photo-generated carriers, and the subsequent interaction is possible between the generated holes on the photocatalytic surface and the membrane lipids of a bacteria. Alternatively, the indirect generation of OH can execute the chemical transformation on the biomaterials [1, 13, 19]. However, these reactions are possible only under light. Based on a study, the coatings of Ag NPs/TiO_2 have been found to achieve antibacterial impact at night [20]. The exceptional bactericidal effect of these coatings is based on the electron storing proficiency of silver NPs that execute the oxidation reactions on the bacterial walls even in the dark [20]. The possible mechanisms of the bactericidal effects of Ag NPs include (i) Ag NPs that adhere with the bacterial surface to change the membrane permeability; (ii) subsequently, silver nanoparticles enter the bacterial cells and cause damage to the DNA; (iii) and then the Ag^+ ions which possess antimicrobial activity [21]. The bactericidal activity of Ag is rather complex and it needs a definite analysis and understanding [1].

Au NPs–Cu_2O can be used as a dual-functional catalyst, which also shows bactericidal activity against *E.coli*, both in the light and in the dark [22]. Additionally, the study of the nanocomposite TiON/PdO for photocatalysts decontamination give activity in low light circumstances [23]. In this nanocomposite, the PdO serves as electron storage and hence, bactericidal effect in dark conditions, while during daytime, TiON is useful for its bactericidal property. Similar mechanism of action is also observed for the anti-bacterial activity of TiO_2/Cu_2O against *E. coli*, titanium oxide against *E. coli* and *E. faecalis* and SnO_2 NPs–Cu_2O nanocubes against *staphylococcus aureus* [24–26]. Table 10.1 lists the use of nanomaterials for biomedical applications.

Table 10.1: Photocatalysts used for biomedical applications.

Photocatalysts	Mechanism of action	Applications	References
TiO_2-nickel hydroxide	Storing holes	Degradation of aldehydes, phenols and alcohols	[17]
TiO_2–nickel hydroxide	Storing holes	Oxidation of formaldehydes and methanol	[27]
TiON/PdO	Storage of electrons	Degradation of methylene blue	[28]
WO_3/TiO_2	Storage of electrons	Degradation of methyl orange	[29]
C_3N_4/CNTs/graphene	Storage of electrons	Removal of phenol	[18]
Titanium dioxide–WO_3	Storage of electrons	Degradation of *E. coli*	[13]
Ag NPs/TiO_2 coatings	Storage of electrons	Disinfection (*S. aureus* and *E. coli*)	[13]
TiO_2 nanosheets	Electron storage	Disinfection (*E. faecalis* and *E. coli*)	[30]
TiO_2/Cu_2O	Electron storage	Disinfection (*E. coli*)	[24]
SnO_2/Cu_2O	Storage of electrons	Disinfection (*S. aureus*)	[25]
TNT–C	Hole storage	Disinfection (*E. coli* and *S. aureus*)	[31]
TiO_2–Ag	Storage of electrons	Deposition process	[32]

10.2.4 Photothermal therapy (PTT) and photodynamic therapy (PDT)

PDT refers to the reactions of cells with light, oxygen and a chemical reagent. PDT is an alternative option for oncologic intrusion, mediated through the tumor cell oxidation and ablation. Based on several investigations and researches, PDT is shown to serve as an exciting option of phototherapy [7]. The mechanism of this therapy is based on the fact that light can excite the tumor-localizing photosensitizers to a higher excited state at a certain wavelength, and that photosensitizer can cross the intersystem to become a triplet-excited photosensitizer. These excited photosensitizers can activate type 1 and 2 PDT reactions for the generation of free radicals and the singlet oxygen, respectively [6, 7].

The mechanism of both of these reactions, PDT and PTT, are somewhat dissimilar but their oxidized products possess related cytotoxic actions. PDT therapy offers several advantages as compared to conventional chemotherapy, such as light-responsive materials are nontoxic and hence can be temporospatial. On the other hand, chemotherapeutics can cause systemic toxicity to the adjacent healthy cells. Moreover, photosensitizers are reusable and hence ultimately improve the quality of life of patients with less invasiveness [33]. Several photosensitizers are approved by FDA for the PDT therapy for several cancers. For example, porfimer sodium is used for the treatment of esophageal cancer, Barrett's esophagus and endobronchial cancer [33]. The surface modifications of the natural biomolecules or polymers allow PDT scheme efficiency for the targeted cancer treatment. For example, MoS_2 is the popular biomaterial used due to its surface modification capability [34].

The PTT therapy has been improved by incorporating inorganic nanomaterials. Based on recent research, the PTT based on gold nanoparticles (Au-NPs) has been shown promising results for the treatment of prostate cancer [35]. Moreover, PTT can be combined with radiation and surgery as a novel synergistic therapy. Additionally, the inorganic nanomaterials can be modified non-covalently and covalently with functional biomolecules due to the large surface-to-volume ratio. These modifications can improve the permeability and retention of photosensitizers in tumors, ultimately minimizing the adverse effects of heat treatment on the adjacent healthy tissues [35, 36]. Carbon- and gold-based materials have also been investigated for PTT therapy; however, these materials possess a weak fluorescence that limits their applications [37].

10.2.5 Optical modulation of cells

Optical modulation of cells, particularly neurons and cardiac cells is another biomedical application of light-responsive biomaterials. Inorganic nanomaterials possess the capability of modulating a single cell using controlled-light platforms via photoelectric or photothermal stimulation [38, 39]. In this way, cellular processes can be detected precisely or can be triggered on demand [40]. This extracellular stimulation of cells, such as the skeletal muscle cells, cardiomyocytes neurons and neurons, is exceptional for the treatment of several disorders [39]. Several functionalized inorganic nanomaterials have been successfully investigated for delivery and dispersion, in the form of drugs, to the targeted organs, such as heart tissues or deep brain [41]. Two main types of cues, electrochemical and temperature, are widely used to control the cell behavior. The photothermal potential of inorganic nanomaterials and their capability to apply thermal stimulation have been well studied to observe the small dynamic changes of temperature on cellular functions [42]. Hence, the photovoltaic and/or photoelectrochemical stimulation can be an alternative choice of studying the non-genetic cell modulations that is associated with less toxic effects and being more sensitive to cells [6, 43].

Semiconductor nanomaterials and quantum dots (QDs) have been investigated for stimulating the cells that are electrically triggered [44]. The light-responsive nanomaterials offer some mechanisms to understand how thermal and/or electrical stimulation can control the protein signaling pathways and gene transcription [45]. Based on such observations, it is manifested that light-responsive nanomaterials can be used as a less incursive and specific treatment of electrophysiological diseases [46].

10.2.6 Drug delivery system

Micro-nanomotors are micro and tiny devices that enable the definite manipulation of substances at the cellular scale [47]. Currently, the light-directed nanovehicles have emerged. They are designed to achieve on-demand and active delivery systems of the drugs, biomolecules and transport cells to the targeted organs or tissues. Owing to their mini size, these devices possess the potential of penetrating into the deep cells to achieve intracellular transport of substances. However, the biocompatibility issue of these devices is the major concern that restricts the selection of appropriate supporting chemicals [48]. Hydrogen peroxide is the commonly employed supporting chemical to drive the nanovehicles; however, this chemical has limited biomedical application. Alternative options are the use of electrical fields, magnetic fields and ultrasound to expel nanovehicles [49].

To combat these issues, the light-assisted nanovehicles are the exceptional option to achieve active delivery of substances or biomolecules in the physiological environment. By controlling the intensity of light, polarization and frequency, this method can provide the freedom to maneuver the light responsive materials [50]. Thus, heating agents or photoelectrochemical cells can absorb the photon energy in the physiological environ and generate the temperature or electrochemical gradient, respectively. Photocatalytic substances, such as Au/Fe_2O_3, Au/Cu_2O and Au/TiO_2 nanomaterials, can undergo the generation of self-electro-osmosis flow through electron–hole separation by a low power of UV irradiation [6]. The use of photothermal nanomaterials is another alternative for the design of light-guided nanovehicles, as these substances can produce local temperature gradients. Among such substances, gold is the most popular material. It can be sputtered or coated on any nanoparticle, such as TiO_2, Fe_2O_3 and mesoporous silica. In general, these light-guided nanovehicles are of a novel technology that combines the knowledge of PTT, PDT, nanoarchitecture and nanofabrication, and these devices assist in the precise cellular drug delivery and in microsurgery [6, 46].

10.3 Application of biomaterials in regenerative medicine and cellular biology

Another biomedical application of light-responsive biomaterials is their use to understand tissue engineering and cellular biology [51]. These biomolecules can induce alterations in the function or heterogeneous structures of natural cellular microenvironment. Some investigations also emphasize the use of light-triggered activation and modulation of tissues for the regeneration of the implanted biomaterials [52]. In view of all of such investigations and studies, the development of novel light-responsive biomaterials can provide exceptional translational clinical applications in future [6]. Table 10.2 lists the light-responsive biomaterials with their biomedical applications, advantages and disadvantages.

Table 10.2: Light-guided biomaterials along with their biomedical applications.

Light-guided biomaterials	Biomedical applications	Applications	Limitations
Graphene, graphene oxide, nanodiamonds and carbon QDs	Antibacterial, PTT, light-guided drug delivery system and cardiac or neuronal cell stimulation	Biocompatibility, chemical inertness and efficient surface area	Poor solubility in physiological environment and nontunable structures
Gold nanoparticles and nanorods	Antibacterial, PTT, light-guided drug delivery system and cardiac or neuronal cell stimulation	Simple functionalization, chemical inertness, biocompatibility and tunable sizes	Limited photodynamic activity and limited drug-loading capacity
$g\text{-}C_3N_4$	Antibacterial and neuron stimulation	Less toxicity, chemical inertness and ease of preparation	Irresponsive to long-wavelength light
TiO_2, ZnO, SnO_2 and ZrO_2	Antibacterial, PDT, PTT, light-guided nanovehicle and light-guided drug delivery	Tunable bandgap, large surface area and economical	Toxic degradation products and limited functionalization
Boron nanosheet and black phosphorus	PTT and light-guided drug delivery	Inherent degradability, biocompatibility and large surface area	Weak photodynamic properties

10.4 Conclusion

Light-responsive materials offer an efficient, feasible and precise platform for controlling the biochemical and biophysical properties of dynamic bionanomaterials. They offer several biomedical applications including the removal of heavy metals from water, pollutant degradation, disinfection, PDT, PTT, drug delivery systems and light-guided devices. To efficiently use solar light, photocatalytic substances are required for the effective conversion of solar light into energy. Recent research has also shifted toward the development of RTCPs that can work both in the presence and in the absence of light. These insights, in turn, emphasize a huge interest in the field of research and investigation into light-responsive nanomaterials and this holds a promising scope for future biomedical revolution.

References

[1] Cai, T., et al., *Recent advances in round-the-clock photocatalytic system: Mechanisms, characterization techniques and applications.* J. Photochem. Photobiol. C: Photochem. Rev., 2019. **39**: p. 58–75.

[2] Fujishima, A. and K. Honda, *Electrochemical Photolysis of Water at a Semiconductor Electrode.* Nature, 1972. **238**(5358): p. 37–38.

[3] Bai, S., et al., *ChemInform Abstract: Steering Charge Kinetics in Photocatalysis: Intersection of Materials Syntheses, Characterization Techniques and Theoretical Simulations.* Chem. Soc. Rev., 2015. **44**.

[4] Linsebigler, A.L., G. Lu, and J.T. Yates, *Photocatalysis on TiO2 Surfaces: Principles, Mechanisms, and Selected Results.* Chem Rev, 1995. **95**(3): p. 735–758.

[5] Schneider, J., et al., *Understanding TiO2 Photocatalysis: Mechanisms and Materials.* Chem Rev, 2014. **114**(19): p. 9919–9986.

[6] Lee, H.P. and A.K. Gaharwar, *Light-Responsive Inorganic Biomaterials for Biomedical Applications.* Adv. Sci., 2020. **7**(17): p. 2000863–2000863.

[7] Dolmans, D.E., D. Fukumura, and R.K. Jain, *Photodynamic therapy for cancer.* Nat Rev Cancer, 2003. **3**(5): p. 380–7.

[8] Zhang, L., et al., *Structure and Properties of Li_2O_3-TiO_2 Nanocomposite Films for Biomedical Applications.* Bioinorg Chem Appl, 2011. **2011**: p. 853048.

[9] Zhou, Q., et al., *Long afterglow phosphor driven round-the-clock g-C3N4 photocatalyst.* J. Photochem. Photobiol. C: Photochem. Rev., 2016. **328**: p. 182–188.

[10] Sakar, M., et al., *Materials and Mechanisms of Photo-Assisted Chemical Reactions under Light and Dark Conditions: Can Day–Night Photocatalysis Be Achieved?* ChemSusChem, 2018. **11**(5): p. 809–820.

[11] Li, Q., et al., *Palladium Oxide Nanoparticles on Nitrogen-Doped Titanium Oxide: Accelerated Photocatalytic Disinfection and Post-Illumination Catalytic "Memory".* Adv. Mater, 2008. **20**(19): p. 3717–3723.

[12] Zhao, D., et al., *Photoinduced Electron Storage in WO3/TiO2 Nanohybrid Material in the Presence of Oxygen and Postirradiated Reduction of Heavy Metal Ions.* J. Phys. Chem. C, 2009. **113**(30): p. 13160–13165.

[13] Tatsuma, T., et al., *Bactericidal effect of an energy storage TiO2–WO3 photocatalyst in dark.* Electrochem. commun., 2003. **5**(9): p. 793–796.

[14] Kasap, H., et al., *Solar-Driven Reduction of Aqueous Protons Coupled to Selective Alcohol Oxidation with a Carbon Nitride–Molecular Ni Catalyst System.* J. Am. Chem. Soc., 2016. **138**(29): p. 9183–9192.

[15] Zhao, D., et al., *Photoinduced Electron Storage in WO3/TiO2 Nanohybrid Material in the Presence of Oxygen and Postirradiated Reduction of Heavy Metal Ions.* J. Phys. Chem., 2009. **113**(30): p. 13160–13165.

[16] Choi, Y., et al., *Sequential Process Combination of Photocatalytic Oxidation and Dark Reduction for the Removal of Organic Pollutants and Cr(VI) using Ag/TiO2.* Environ. Sci. Technol., 2017. **51**(7): p. 3973–3981.

[17] Takahashi, Y. and T. Tatsuma, *Oxidative Energy Storage Ability of a TiO2–Ni(OH)2 Bilayer Photocatalyst.* Langmuir, 2005. **21**(26): p. 12357–12361.

[18] Zhang, Q., et al., *Metal-Free Photocatalyst with Visible-Light-Driven Post-Illumination Catalytic Memory.* ACS Appl. Mater. Interfaces, 2017. **9**(26): p. 21738–21746.

[19] Sunada, K., et al., *Bactericidal and Detoxification Effects of TiO2 Thin Film Photocatalysts.* Environmental Science & Technology, 1998. **32**(5): p. 726–728.

[20] Cao, H., et al., *Electron storage mediated dark antibacterial action of bound silver nanoparticles: Smaller is not always better.* Acta Biomaterialia, 2013. **9**(2): p. 5100–5110.

[21] Li, Q., et al., *Antimicrobial nanomaterials for water disinfection and microbial control: Potential applications and implications.* Water Res., 2008. **42**(18): p. 4591–4602.

[22] Kuo, M.-Y., et al., *Au@Cu2O core@shell nanocrystals as dual-functional catalysts for sustainable environmental applications.* Appl. Catal. B, 2019. **242**: p. 499–506.

[23] Li, Q., et al., *Memory antibacterial effect from photoelectron transfer between nanoparticles and visible light photocatalyst.* J. Mater. Chem., 2010. **20**: p. 1068–1072.

[24] Liu, L., et al., *Synthesis of Cu2O Nanospheres Decorated with TiO2 Nanoislands, Their Enhanced Photoactivity and Stability under Visible Light Illumination, and Their Post-illumination Catalytic Memory.* ACS Appl. Mater. Interfaces, 2014. **6**(8): p. 5629–5639.

[25] Liu, L., et al., *Post-illumination activity of SnO2 nanoparticle-decorated Cu2O nanocubes by H2O2 production in dark from photocatalytic "memory".* Sci Rep, 2016. **6**: p. 20878.

[26] Wang, G., et al., *Ultrathin titanium oxide nanosheets film with memory bactericidal activity.* Nanoscale, 2016. **8**(42): p. 18050–18056.

[27] Yang, F., et al., *Oxidation of methanol and formaldehyde to CO2 by a photocatalyst with an energy storage ability.* PCCP, 2010. **12**: p. 5166–70.

[28] Li, Q., et al., *Palladium Oxide Nanoparticles on Nitrogen-Doped Titanium Oxide: Accelerated Photocatalytic Disinfection and Post-Illumination Catalytic "Memory".* Adv. Mater , 2008. **20**: p. 3717–3723.

[29] Li, Y., et al., *Low-temperature electrodeposited Co-doped ZnO nanorods with enhanced ethanol and CO sensing properties.* Sens. Actuators B Chem., 2012. **161**(1): p. 734–739.

[30] Wang, G., et al., *Ultrathin titanium oxide nanosheets film with memory bactericidal activity.* Nanoscale, 2016. **8**.

[31] Wang, G., et al., *An antibacterial platform based on capacitive carbon-doped TiO2 nanotubes after direct or alternating current charging.* Nat Commun, 2018. **9**(1): p. 2055.

[32] Cai, T., et al., *"Dark Deposition" of Ag Nanoparticles on TiO2: Improvement of Electron Storage Capacity To Boost "Memory Catalysis" Activity.* ACS Appl. Mater. Interfaces, 2018. **10**(30): p. 25350–25359.

[33] Allison, R.R. and C.H. Sibata, *Oncologic photodynamic therapy photosensitizers: a clinical review.* Photodiagnosis Photodyn Ther, 2010. **7**(2): p. 61–75.

[34] Liang, J. and B. Liu, *ROS-responsive drug delivery systems*. Bioeng Transl Med, 2016. **1**(3): p. 239–251.
[35] Stern, J.M., et al., *Initial evaluation of the safety of nanoshell-directed photothermal therapy in the treatment of prostate disease*. Int J Toxicol, 2016. **35**(1): p. 38–46.
[36] Deokar, A.R., et al., *Graphene-based "hot plate" for the capture and destruction of the Herpes simplex virus type 1*. Bioconjugate Chem., 2017. **28**(4): p. 1115–1122.
[37] Sun, X., et al., *NIR-triggered thermo-responsive biodegradable hydrogel with combination of photothermal and thermodynamic therapy for hypoxic tumor*. Asian J Pharm Sci, 2020. **15**(6): p. 713–727.
[38] Shah, S., A. Solanki, and K.B. Lee, *Nanotechnology-Based Approaches for Guiding Neural Regeneration*. Acc Chem Res, 2016. **49**(1): p. 17–26.
[39] Li, J., J. Liu, and C. Chen, *Remote Control and Modulation of Cellular Events by Plasmonic Gold Nanoparticles: Implications and Opportunities for Biomedical Applications*. ACS Nano, 2017. **11**(3): p. 2403–2409.
[40] Parameswaran, R., et al., *Photoelectrochemical modulation of neuronal activity with free-standing coaxial silicon nanowires*. Nat Nanotechnol, 2018. **13**(3): p. 260–266.
[41] Parker, S.G., et al., *A photoelectrochemical platform for the capture and release of rare single cells*. Nat Commun, 2018. **9**(1): p. 2288.
[42] Yoo, S., et al., *Photothermal inhibition of neural activity with near-infrared-sensitive nanotransducers*. ACS Nano, 2014. **8**(8): p. 8040–9.
[43] Marino, A., et al., *Gold Nanoshell-Mediated Remote Myotube Activation*. ACS Nano, 2017. **11**(3): p. 2494–2508.
[44] Rotenberg, M.Y. and B. Tian, *Talking to cells: semiconductor nanomaterials at the cellular interface*. Advanced biosystems, 2018. **2**(4): p. 1700242.
[45] Huang, K., Q. Dou, and X.J. Loh, *Nanomaterial mediated optogenetics: Opportunities and challenges*. RSC Adv., 2016. **6**.
[46] Choi, S., et al., *Recent Advances in Flexible and Stretchable Bio-Electronic Devices Integrated with Nanomaterials*. Adv Mater, 2016. **28**(22): p. 4203–18.
[47] Peng, F., Y. Tu, and D. Wilson, *Micro/nanomotors towards: In vivo application: Cell, tissue and biofluid*. Chem. Soc. Rev., 2017. **46**.
[48] Gao, W., S. Sattayasamitsathit, and J. Wang, *Catalytically propelled micro-/nanomotors: how fast can they move?* Chem. Rec., 2012. **12**(1): p. 224–231.
[49] Restrepo-Pérez, L., et al., *Biofunctionalized self-propelled micromotors as an alternative on-chip concentrating system*. Lab Chip, 2014: p. DOI: 10.1039/C4LC00439F.
[50] Wang, J., et al., *Light-Driven Micro/Nanomotor for Promising Biomedical Tools: Principle, Challenge, and Prospect*. Acc. Chem. Res., 2018. **51**(9): p. 1957–1965.
[51] Ruskowitz, E.R. and C.A. DeForest, *Photoresponsive biomaterials for targeted drug delivery and 4D cell culture*. Nat. Rev. Mater., 2018. **3**(2): p. 1–17.
[52] Lee, T.T., et al., *Light-triggered in vivo activation of adhesive peptides regulates cell adhesion, inflammation and vascularization of biomaterials*. Nat Mater, 2015. **14**(3): p. 352–60.

Index

https://doi.org/10.1515/9783110768749-011